Microscale Chemistry Laboratory Manual

Student's Edition

ALAN SLATER

Stratford Central Secondary School
Stratford, Ontario

GEOFF RAYNER-CANHAM

Sir Wilfred Grenfell College
Corner Brook, Newfoundland

Addison-Wesley Publishers Limited
Don Mills, Ontario • Reading, Massachusetts
Menlo Park, California • New York
Wokingham, England • Amsterdam
Bonn • Sydney • Singapore
Tokyo • Madrid • San Juan

Reviewer
Ed Doadt
Cameron Heights Collegiate
Kitchener, Ontario

Editorial
Lesley Haynes
Louise MacKenzie

Photo cover courtesy of Boreal Laboratories Ltd.

Canadian Cataloguing in Publication Data
Slater, Alan, 1942—
Microscale chemistry laboratory manual

ISBN 0-201-60215-6

1. Microchemistry — Laboratory manuals.
I. Rayner-Canham, Geoffrey. II. Title.

QD45.S53 1994 543'.0813 C93-095125-5

ISBN 0-201-60215-6
This book contains reycled paper and is acid free.
Printed and bound in Canada.

A B C D E F — MP — 99 98 97 96 95 94

Table of Contents

Introduction

Chemistry is an experimental science. It is not a subject just to read about in textbooks; it is an exploration of how the world works. You must DO and SEE chemistry. The experiments in this laboratory manual serve this purpose. There are a number of unique features in this manual.

THE MICROSCALE APPROACH

The experiments in this laboratory manual are conducted using the new techniques of microscale experimentation. All experiments are performed with very small quantities of solutions and solids. This approach has a number of advantages over the traditional beaker-and-test-tube methods:

More experiments

- Microscale experiments can be completed more quickly. You can do more hands-on chemistry in the same amount of lab time that it would take to do one experiment using traditional methods.

Safer

- The experiments are much safer because the quantities of any hazardous materials are so small.
- There is less chance of accidents occurring because you are using plastic containers instead of the traditional glass containers.

Better for the Environment

- There is less material to dispose of because of the small quantities used, so damage to the environment is reduced.

Lower Cost

- By using small amounts of chemicals and inexpensive plastic equipment, your school can use the money it saves on other equipment and materials to further your education in chemistry and other subjects.

SCIENCE, TECHNOLOGY, AND SOCIETY ISSUES

Chemistry is not just what happens in the chemistry laboratory. Chemistry happens all around you. Each essay on an aspect of chemistry has links to an experiment or a set of experiments. This illustrates that the principles that you study in the lab really do have relevance to your daily life.

EXTENSIONS

Real chemistry is not repeating what other people have done; it involves performing experiments that you design. Many experiments have extensions to give you the freedom to explore beyond the known and expected.

ENVIRONMENTAL APPLICATIONS

Some of the experiments raise some very important issues concerning the environment. We have added discussion topics to explore these problems.

We believe chemistry is one of the most interesting fields to study and we hope that when you experience the fascination of chemistry in the laboratory, you will agree.

Alan Slater and Geoff Rayner-Canham

Safety in the Microscale Laboratory

Because you use very small quantities of chemical substances and most of the equipment is plastic rather than glass, the microscale laboratory experiments you perform should be safe. Nevertheless, some of the chemicals and procedures that you will be using do present small but significant hazards. For this reason, it is important to follow the safety recommendations listed for each experiment. The safety icons and their meanings, listed below, are used throughout the manual.

Eye Hazard

Wear eye protection.

Corrosive Substance Hazard

One or more of the substances are corrosive. Immediately wash off any spills on your skin and clothing.

Fire Hazard

One or more of the substances used are flammable OR an open flame is used in the experiment. Tie back long hair and fasten loose clothing.

Toxic Substance Hazard

One or more of the substances are toxic and/or carcinogenic. Avoid contact with these substances. Wash your hands thoroughly after the experiment.

Inhalation Hazard

Avoid inhaling these substances.

Thermal Burn Hazard

Do not touch hot equipment.

Disposal Hazard

Dispose of this chemical only as directed by your teacher.

Electric Shock Hazard

Be careful when using electrical equipment.

What Is WHMIS?

The Workplace Hazardous Materials Information System, or WHMIS, is a national system. It is designed to ensure that all employers obtain the information they need to inform and train their employees about the safe handling and usage of hazardous materials. Through legislation, the hazards of materials produced or sold in, imported into, or used within workplaces in Canada are identified by standard classification criteria. The goal of WHMIS is to reduce the incidence of illnesses and injuries resulting from the use of hazardous materials in the workplace.

Suppliers of controlled products must convey hazard information to purchasers in a specified manner. This can be labelling on the controlled products or containers of the controlled product in the form of a Material Safety Data Sheet (MSDS).

Employers are required to develop appropriate workplace labelling and other forms of warning about controlled products produced in their workplace processes, make MSDSs available to their workers, and provide for worker education on the safe use of hazardous materials.

Provided courtesy of Alberta Occupational Health and Safety

Microscale Laboratory Equipment

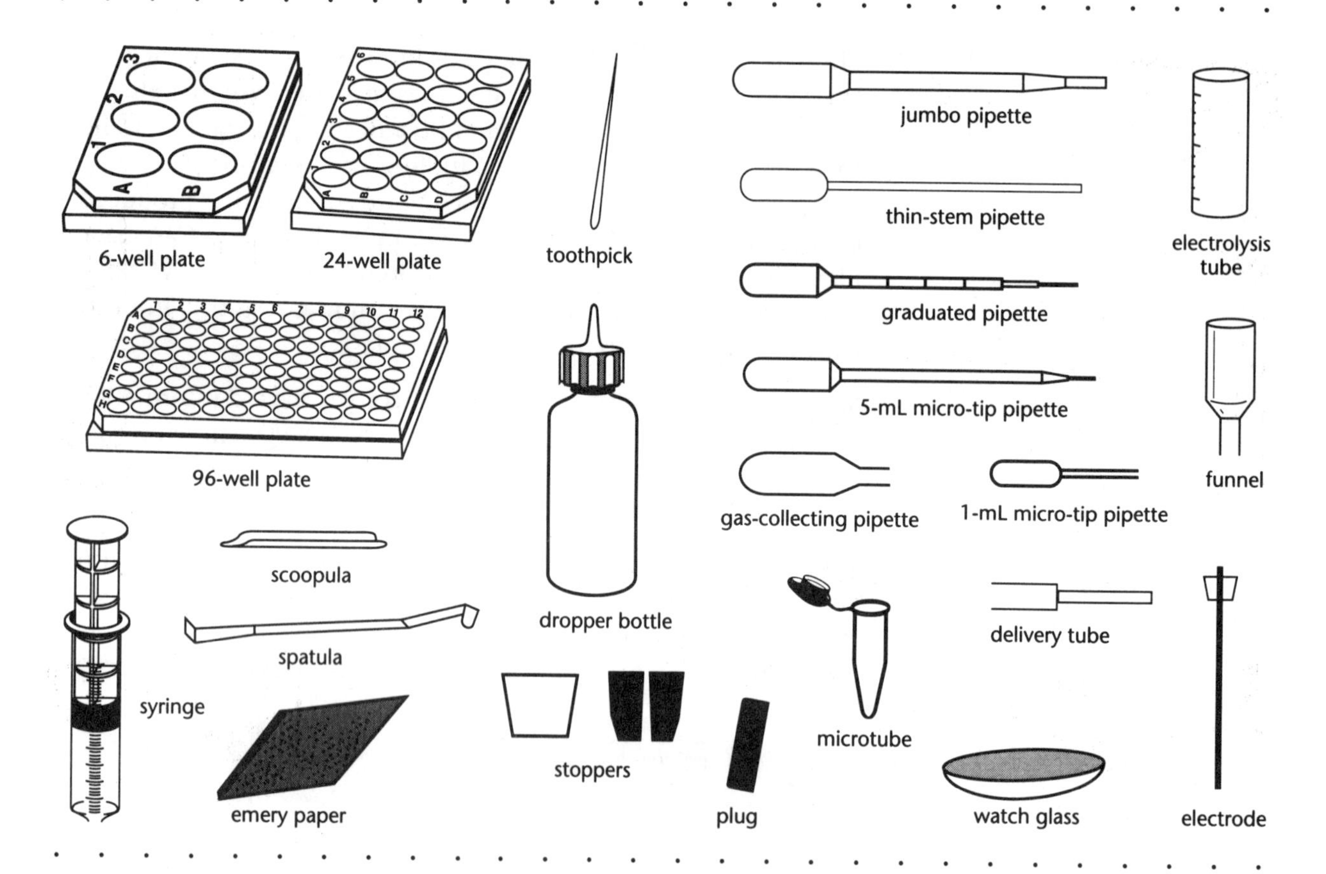

Balance: used for weighing solids; should measure to at least a hundredth of a gram.

9-V Battery: used to power conductivity tester and various electrolysis reactions.

Beaker: glass or preferably plastic; mainly used as a water bath.

Burner: used as a heat source mainly for testing certain gases.

Conductivity tester: used to check the electrical conductivity of various materials.

Cotton swab: mainly used to clean well plates.

Delivery tube: plastic; made from the cut off tip of a graduated pipette. Used with a stopper to collect gases from a 24-well plate.

Dropper bottle: plastic or glass; various models; used as a container for most solutions; a 30-mL bottle is large enough for most experiments.

Electrode: made from a 00 stopper and pencil lead or copper wire.

Electrolysis tube: plastic; made from a graduated serological pipette; used to collect and store the products of electrolysis.

Emery paper: used to clean metals.

Funnel: flexible plastic; made from a graduated pipette; used to place solids into cut-off jumbo pipettes.

Gas-collecting pipette: flexible plastic; made from a cut-off graduated pipette; used to collect gas samples.

Graduated cylinder: plastic or glass; 10-mL size is used to measure large volumes of solutions.

Graduated pipette: flexible plastic; dispenses 25 drops/mL; graduated in 0.25 increments to 1.0 mL; used to measure small volumes of solution; also used to make gas-collecting pipettes.

Jumbo pipette: flexible plastic; dispenses 22 drops/mL; graduated in 0.50 increments to 3 mL; used to dilute larger amounts of solutions; when its tip is cut off, it is used as a test tube.

1-mL micro-tip pipette: flexible plastic; dispenses 43 drops/mL; used to dispense small drops of solution.

5-mL micro-tip pipette: flexible plastic; dispenses 50 drops/mL; used in experiments that require larger amounts of solution dispensed in small drops.

Microtube: flexible plastic; graduated in 0.5 mL increments; have plastic caps. Used to make electrolysis tubes; also used as test tubes.

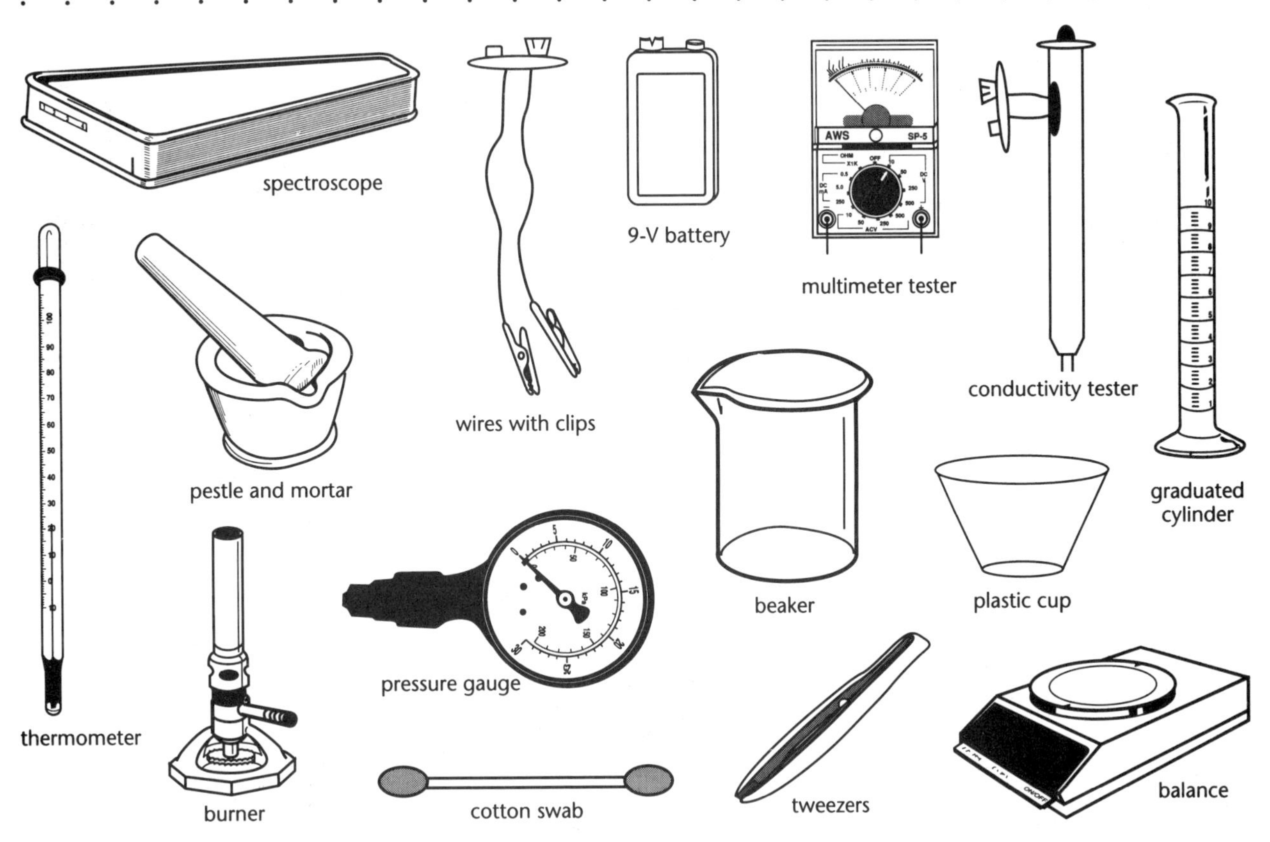

Multimeter tester: used to measure voltage from electrochemical cells and to check for problems in electrolysis circuits.

Pestle and mortar: porcelain; used to grind solids into smaller pieces.

Plastic cup: has a volume of about 25 mL; two used with an elastic band to form a calorimeter.

Plug: plastic; used to seal the tip of a syringe.

Pressure gauge: used to measure pressure of a gas.

Scoopula: metal; used to dispense relatively large amounts of solids.

Spatula: metal; used to dispense relatively small amounts of solids.

Spectroscope: a black box with a diffraction grating used to analyse the colours found in light.

Stoppers: #2 1-hole rubber stopper used to seal wells of, and collect cases from, a 24-well plate; 00 cork stopper used to seal the end of an electrolysis tube and, when pierced with a pencil lead or copper wire, used as an electrode.

Syringe: plastic; various sizes; used to study the properties of gases; with its tip cut off, used as a gas-collecting apparatus.

Thermometer: used to measure temperatures of solutions or water baths.

Thin-stem pipette: flexible plastic; dispenses 25 drops/mL; used to place solutions in a jumbo pipette.

Toothpick: wood; used as a wooden splint to test for gases. Also useful, when wetted, to pick up grains of solid to add to a well; also used as a well stirrer.

Tweezers: metal; used to handle solids that are reactive or that have been in corrosive substances.

Watch glass: glass; used to hold small amounts of solid.

6-well plate: plastic; well volume of 16.8 mL; used to hold large amounts of liquid or solid.

24-well plate: plastic; well volume of 3.4 mL; used in titrations, gas collections, and electrochemical experiments.

96-well plate: plastic; well volume of 0.37 mL; used in many different types of experiment.

Wires with clips: metal; used to connect a battery to electrodes.

The Importance of Chromatography

Plants and animals don't consist of pure substances. They consist of millions of different complex mixtures of chemical substances. The area of chemistry that investigates the chemical compounds found in nature is called natural products chemistry. Natural products chemists could spend thousands of years in the laboratory trying to make every chemical compound possible. Instead, natural products chemists take advantage of the immense range of compounds that nature has produced. Since each plant and animal contains a huge number of different compounds, it's no easy job to separate all the compounds, so natural products chemists make use of an important tool—chromatography.

There are different types of chromatography. The simplest type (which you will be using) is paper chromatography, but all types use the same concept. The mixture of compounds obtained from a sample is dissolved in a solvent. The solvent then travels through a material, such as filter paper, carrying the different compounds in the mixture along with it. These compounds tend to "stick" to the paper and each compound "sticks" to a different degree. Thus the compounds that "stick" the least will travel along with the front edge of the solvent. Those compounds that "stick" the most will hardly move along the paper at all. In this way, we can separate the components of a mixture.

When the components have been separated, the chemists can find out which of them are useful. Many compounds from biological sources are used as medicines: digitalis, a heart-stimulating drug is a compound extracted from the foxglove plant; and a promising new drug for the treatment of arthritis was found in extracts from a sea urchin.

Chromatography won't help natural products chemists find new compounds if the chemists don't have plant and animal samples. Perhaps one of the strongest arguments for protecting the world's tropical rain forests is that they contain a vast number of plant and animal species. These forests are thus a great source of compounds that may provide cures for cancer, AIDS, and other diseases.

RELATED EXPERIMENT

Experiment 1 Chromatography of an Ink

Experiment 1

Chromatography of an Ink

INTRODUCTION

Chromatography is a technique that separates small amounts of a mixture efficiently into its component parts. Paper chromatography involves placing a drop of a coloured mixture on a strip of chromatography paper or filter paper. The strip is then suspended in water which moves up the paper by capillary action. The components in the mixture begin to migrate up the paper at different rates, resulting in their separation.

PROBLEM

Are all black inks made up of the same substances?

APPARATUS AND MATERIALS

Per pair of students:

1 24 well plate
6 jumbo pipettes
1 thin-stem pipette
6 strips of chromatography paper or filter paper (0.5 cm wide)
6 different water-soluble black pens
tap water

PROCEDURE

1. Take a strip of chromatography paper, 0.5 cm by 15 cm. At about 2 cm from one end, draw a pencil line across it.
2. Use one of the water-soluble pens and place a dot on the centre of this line. This dot should be about the size of a period.
3. Insert a thin-stem pipette, with water in it, into the bulb of a jumbo pipette. Place 5 drops of water in the bottom of the jumbo pipette, and then remove the thin-stem pipette.
4. Slide the chromatography paper inside the jumbo pipette, as shown in the diagram on the next page, so that the end of the paper closest to the pencil line is in the water.

Fold the other end of the paper over the rim of the pipette. Be sure not to immerse the dot in the water.

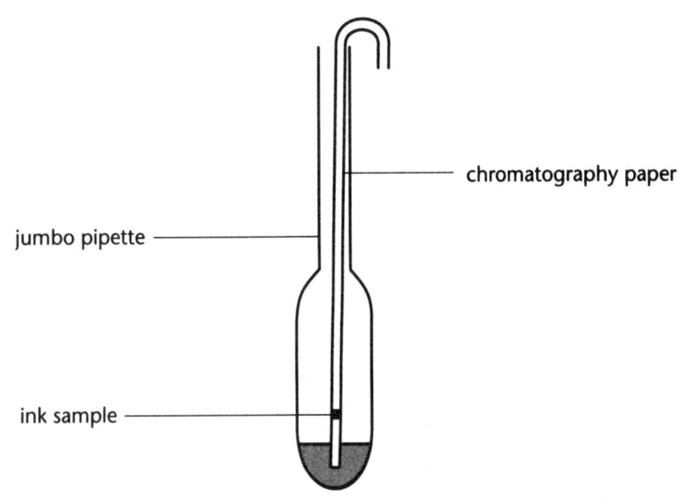

5. Place the jumbo pipette in a well in the 24-well plate.
6. When the water has moved up the paper a distance of about 12 cm, remove the paper and hang it on the outside of the pipette to dry.
7. Repeat Steps 1 to 6 for each of five inks from the pens of other students in the class.

Optional: Take an unknown ink and try to determine its make. Have a friend put the dot on the chromatography paper. Make sure you do not see from which pen it came.

ANALYSIS

1. Did all the inks contain the same dyes?
2. What made the dyes move up the paper?
3. Did some inks not move? Why? What could be done to make them move?
4. This is a qualitative experiment. How could you make it a quantitative experiment?

EXERCISES

1. Scientists use paper chromatography to identify unknown components of a mixture, using R_f values. What is an R_f value?

EXTENSION

1. Use R_f values to find quantitatively the components of an unknown mixture.
2. What effects do other solvents have on the chromatography?

Chemical Changes in the Kitchen

What did you have for breakfast? Did its preparation involve a chemical change? How could you tell if a chemical change occurred? If you had an egg for breakfast, that is easy. Upon cooking, the egg changed from a liquid to a solid. This phase change is evidence of a chemical change. Toasting bread is another example. The surface of the bread becomes harder and darker in colour. A colour change is a good indication of a chemical change. The preparation of other breakfast foods, such as brewing coffee, involves physical changes. In this process, some of the chemical compounds in coffee beans are dissolved in hot water to give the characteristic odour and flavour of coffee. Finally, breakfast ends with a chemical change, that of the reaction between the grease on the plates and the dish-washing detergent. As you eat lunch and dinner look for evidence of chemical changes there, too.

RELATED EXPERIMENT

Experiment 2 Chemical Changes

Experiment 2

Chemical Changes

INTRODUCTION

One of the most important scientific skills is the ability to make observations and record them clearly and completely. When you have made the observations, you can make deductions about what you have seen.

In the chemistry laboratory, you will often observe chemical changes (chemical reactions) and have to deduce what has happened. In this experiment, you will be mixing solutions of pairs of chemical compounds and seeing what evidence there is that chemical reactions have occurred.

PROBLEM

How can you identify a chemical change?

APPARATUS AND MATERIALS

Per pair of students:

1 96-well plate
6 dropper bottles containing solutions labelled A to F

SAFETY

Solution C is corrosive. Avoid contact with skin and clothing. Flush any contacted area with running water.

PROCEDURE

1. Place 1 drop of solution A in each of wells A1 to A5.
2. Add 1 drop of solution B to well A1, 1 drop of solution C to well A2, 1 drop of solution D to well A3, 1 drop of solution E to well A4, and 1 drop of solution F to well A5. Record your observations (if any) in each case.

ANALYSIS

1. What different types of evidence indicate that chemical reactions have occurred?
2. Does a chemical reaction always occur when different substances are mixed?

EXERCISES

1. Are there any other ways of detecting a chemical reaction?

EXTENSION

Cooking food involves chemical reactions. Review a typical day's meals. Explain how you can tell when the desired cooking reaction has occurred for each item of food.

Chemistry Is All around Us

People often think that chemistry only happens in laboratories, but that's not true. It happens all around us—and in us. The human body is a complex chemical "factory" surrounded by a protein bag called skin. Our bodies take the chemical mixtures in our food and convert them into the wide variety of chemicals that we need to survive and grow. The food we eat also provides the energy our bodies need to make all these different chemicals.

Generally, the chemicals in our bodies and in our foods are too complex to discuss in an introductory chemistry course. But there are many simple chemicals you can find around your home. These tend to be cleaning products: sodium hydroxide (oven cleaners and drain openers), ammonia (glass cleaners), and sodium hypochlorite (bleach). All of these chemicals are corrosive and poisonous, as are most cleaning products. Around the home, it is amazing how casually some people handle these chemicals. In the laboratory, people wear coats, safety glasses, and even gloves when they use these same chemicals. At home, few of these precautions are taken. As a result of this chemistry course, we hope you will treat chemicals at home as cautiously as you do in the laboratory.

RELATED EXPERIMENT

Experiment 3 Identification of Household Substances

Experiment 3

Identification of Household Substances

INTRODUCTION

Chemists identify an unknown substance by identifying the chemical reactions that are unique to each of many known substances. In this experiment, you will be given a compound of unknown composition. You will then perform a variety of chemical reactions until you obtain a match with that of a known compound. You will then be able to tell the identity of the unknown compound.

Note: Before conducting this experiment, review Experiment 2 so that you can recognize when a chemical reaction has occurred!

PROBLEM

How can you identify an unknown household substance?

APPARATUS AND MATERIALS

Per pair of students:

1 96-well plate
1 24-well plate
1 toothpick
1 1-mL micro-tip pipette
paper towel
1 spatula
table salt (sodium chloride)
baking soda (sodium bicarbonate)
washing soda (sodium carbonate)
drain opener (sodium hydroxide)
toilet-bowl cleaner (sodium bisulfate)
Epsom salts (magnesium sulfate)
distilled water
Ex-lax solution (phenolphthalein)
vinegar (acetic (ethanoic) acid)
methyl orange solution
calcium chloride (ice-melter)
unknown substance from teacher

SAFETY

Solid sodium hydroxide is very corrosive. Handle with care. Avoid contact with skin and clothing. Flush any contacted area with running water.

PROCEDURE

1. Write an abbreviated name (for example, ts for table salt) for each of the first six household substances from the Apparatus and Materials list on a piece of paper. When a 24-well plate is placed on the paper, each name should appear under one of the first 6 wells in the plate. This well plate will hold your "stock solution" from which you will take samples for testing.

2. Pour an amount of each household substance, equal in volume to half an aspirin tablet, into the well over the matching name.
3. Half-fill each well with distilled water and use a toothpick to stir the contents of each well. Be sure to wipe the toothpick with a paper towel before inserting it in the next well.
4. Using a 1-mL micro-tip pipette, place 5 drops of the solution of each substance into the wells of the 96-well plate as shown below. Rinse the pipette thoroughly before using it for a different solution.

Solution	Well
sodium chloride	A1 to D1
sodium bicarbonate	A2 to D2
sodium carbonate	A3 to D3
sodium hydroxide	A5 to D5
sodium bisulfate	A4 to D4
magnesium sulfate	A6 to D6

5. Using the pipette, add 5 drops of each reagent listed below to the indicated wells. Rinse the pipette thoroughly before using for a different solution. (Column 7 is for your unknown sample.) Record your observations of each chemical reaction in a table.

Solution	Well
phenolphthalein	A1 to A7
vinegar (acetic acid)	B1 to B7
methyl orange	C1 to C7
calcium chloride	D1 to D7

6. Collect an unknown substance from your teacher. Dissolve some of it in distilled water in well A7 of the 24-well plate. Place 5 drops of the resulting solution in wells A7 to D7 of the 96-well plate. The reagents were placed in these wells in Step 5. Compare the results of the reactions in column 7 with those in columns 1 to 6. Which one shows the same results as column 7?

ANALYSIS

1. What different types of evidence indicate that chemical reactions have occurred?
2. What is the identity of the unknown substance? What is your evidence?

EXERCISES

1. What are the limitations of this method of analysis?

EXTENSION

Your teacher will supply some other simple substances that are used as household products. Repeat the above experiment with these substances. If necessary, try to devise some additional tests to identify definitively each substance.

Experiment 4

Changes in Mass during Chemical Reactions

INTRODUCTION

It is almost impossible to observe chemical changes at the atomic level. In the experiments you will perform in this course, you can observe macroscopic (large) chemical changes. From your observations, you can deduce what is happening at the atomic level.

PROBLEM

In a chemical reaction, how does the mass of the reactants compare to the mass of the products?

APPARATUS AND MATERIALS

Per pair of students:

1 24-well plate
1 balance
1 jumbo pipette
2 thin-stem pipettes
dropper bottles of:
    lead(II) nitrate solution
    potassium iodide solution
    barium chloride solution
    sodium sulfate solution

SAFETY

Lead and lead compounds are toxic and can cause birth defects if ingested by pregnant women. Barium compounds are also toxic. However, in the dilute solutions used here, they present little risk. As an extra precaution, wash your hands thoroughly after the experiment.

PROCEDURE

1. Before starting, make sure your pipettes are clean and dry on the outside.
2. Place about 5 drops of lead(II) nitrate solution in the bottom of the jumbo pipette using a thin-stem pipette.
3. Fill a thin-stem pipette with potassium iodide solution. Place this pipette in the jumbo pipette, as shown in the diagram below. Make sure the reactants do not mix at this point.

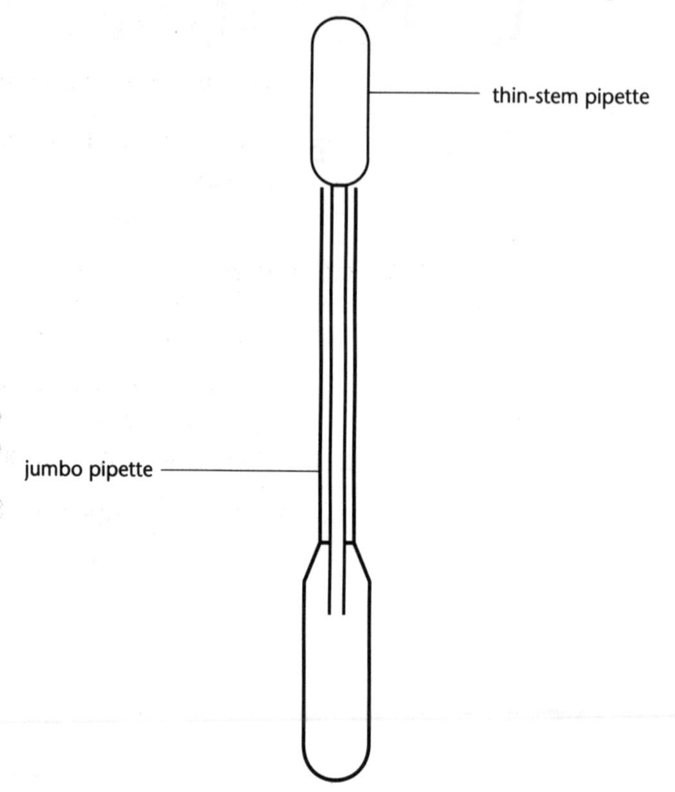

4. Zero the balance with a 24-well plate on it.
5. Place the jumbo pipette containing the thin-stem pipette into a central well of the plate and measure the mass.
6. Squeeze 5 drops of the potassium iodide solution into the jumbo pipette and observe what happens.
7. Measure the mass of the whole assembly again.
8. Clean the pipettes.
9. Repeat Steps 1 to 8, replacing the lead(II) nitrate solution with the barium chloride solution, and the potassium iodide with the sodium sulfate solution.

10. Dispose of the lead solutions as instructed by your teacher.

ANALYSIS

1. Why is it important to make sure the outsides of the pipettes are clean and dry?
2. For each pair of solutions, what evidence suggested there was a chemical reaction?
3. For each pair of solutions, compare the mass before and the mass after their mixing. What conclusion can be drawn?
4. If the reaction produced a gas, what would happen to the mass? Why?
5. What law does this experiment illustrate?
6. As a tree grows, it gets taller. Does this contradict the law? Explain.

EXTENSION

1. How would you do this experiment for a reaction that produces a gas which you want to collect? For example, consider the reaction of vinegar with sodium bicarbonate (baking soda). You might need a resealable plastic bag.
2. Write a word equation for each reaction.

Chemistry Discoveries

There are some fundamental laws that are vital to our understanding of chemistry. These include the Law of Conservation of Mass and the Law of Definite Proportions. Chemistry is a cumulative body of knowledge and chemists rely on these laws to help them understand chemical changes. Yet, it is easy for people to have the false view of science (and chemistry in particular) that everything is known and everything follows rigid laws.

The excitement of science comes from the fact that there are always unexpected discoveries being made. In the 1930s, an American chemist, Dr. Roy Plunkett, was studying a gas named tetrafluoroethylene. One day, he found a cylinder containing the gas that seemed to be empty, even though the day before it had been full. Most people would have thought that the gas had leaked out overnight. Plunkett, however, had the cylinder cut open and he found a white solid inside. Intrigued, he studied the substance and found it was exceptionally slippery. Some rust in the cylinder had caused the gas to form this new compound and this reaction could be made to happen quite easily. Why was this unexpected discovery so important? Well, that white substance is what is now known as Teflon, which coats our pots and pans and has a host of other uses.

Many discoveries are made by prediction but until a chemist goes into the laboratory and actually tries the experiment, he or she cannot be sure what the outcome will be. For example, Stephanie Kwolek and her colleagues at Du Pont research laboratories deduced that it was possible to synthesize a completely new type of plastic fibre that would be much stronger and have a much higher melting point than traditional fibres, such as nylon and polyester. After many unsuccessful experiments, they finally produced such a material, which they named Kevlar. This chemical compound is now used to manufacture bullet-proof vests. The lives of police officers around the world have been saved by their wearing this low-density material that is flexible, yet many times stronger than the same thickness of steel thread.

RELATED EXPERIMENT

Experiment 5 Law of Definite Proportions

Experiment 5

Law of Definite Proportions

INTRODUCTION

One of the fundamental ideas in science is that chemicals react in precise ratios. In this experiment, you will perform two separate chemical reactions to investigate the ratio in which certain ions combine. You will use solutions that each contain the same number of moles of ions per unit volume.

PROBLEM

What are the common ratios in which ions combine?

APPARATUS AND MATERIALS

Per pair of students:

1 96-well plate
3 1-mL micro-tip pipettes and labels
2 toothpicks
silver ion solution
chloride ion solution
chromate ion solution

SAFETY

Chromate ions are suspected carcinogens. However, in the dilute solution used here, they present little risk. As an extra precaution, wash your hands thoroughly after the experiment. Silver ions stain skin and clothing.

PROCEDURE

1. Half-fill the pipettes with one of the silver, chloride, and chromate ion solutions. Label the pipettes.

2. Put drops of the silver ion solution and chloride ion solution in the wells as indicated below. Be sure to hold the pipette vertically while adding the drops to ensure a consistent drop size.

	A1	A2	A3	A4	A5	A6	A7	A8	A9	A10	A11
Ag^+	1	2	3	4	5	6	7	8	9	10	11
Cl^-	11	10	9	8	7	6	5	4	3	2	1

3. The wells should be filled to equal depths. Stir the contents of each well with a toothpick. Wipe the toothpick with paper towel between each stirring.

4. Repeat Steps 2 and 3. Use row H in the well plate, putting drops of the silver ion solution and chromate ion solution in the wells as indicated below.

	H1	H2	H3	H4	H5	H6	H7	H8	H9	H10	H11
Ag^+	1	2	3	4	5	6	7	8	9	10	11
CrO_4^{2-}	11	10	9	8	7	6	5	4	3	2	1

5. Allow the precipitates to settle for about 5 min. Record your observations.

6. After the precipitates have settled, note which well in each row appears to contain the most precipitate.

ANALYSIS

1. (a) Which well in row A gave the most precipitate?
   (b) What is the ratio of silver ion to chloride ion in this well?
   (c) From looking at the charges on the ions, would you expect this ratio? Explain.

2. (a) Which well in row H gave the most precipitate?
   (b) What is the ratio of silver ion to chromate ion in this well?
   (c) From looking at the charges on the ions, would you expect this ratio? Explain.
   (d) What evidence do you have that the chromate is in excess in some wells?

EXTENSION

Using the method in this experiment, try other combinations of ions that give insoluble compounds. Determine the ratio of the ions in the same way. Did you find any other ion ratios?

Experiment 6

Flame Spectra

INTRODUCTION

You may have seen the spectra of different gases. Each gas has a unique "fingerprint," which is only seen when energy (in the form of electricity in this case) is passed through the gas.

This experiment demonstrates how we see the "fingerprints" of solids and get the range of colours produced when they are heated.

PROBLEM

Will different atoms produce the same spectrum?

APPARATUS AND MATERIALS

Per pair of students:

1 burner
1 spectroscope
small plastic containers of:
- copper(II) sulfate
- potassium bicarbonate
- barium hydroxide
- lithium hydroxide
- strontium chloride
- unknown substance

SAFETY

Barium compounds are toxic. Do not squirt any dust into the air. Wash your hands thoroughly after the experiment. Barium and lithium hydroxides are corrosive. Avoid contact with skin and clothing. Flush any contacted area with running water.

PROCEDURE

1. Light your burner and keep the air intake valve open.
2. Take one of the plastic containers and shake it to create a dust. Remove the cap and hold the opening of the container close to the air intake valve of your burner. Observe the colour of the flame and then view the spectrum through the spectroscope. Record in a table the main lines that you see through the spectroscope.
3. Repeat Step 2 for each of the other powders.
4. From your teacher, get a powder whose identity is unknown. Repeat Step 2. Observe the colour of the flame and then view the spectrum through the spectroscope. Identify the substance.

ANALYSIS

1. From your observations write a brief account to justify the identity of the powder that caused the colours you observed in Step 4.
2. (a) How does the burner's flame cause the chemical substance to produce light?
   (b) Where, in the atom, is this light produced?

EXTENSION

1. How could an astronomer determine the components of a distant star?
2. Does an atom emit light outside the visible spectrum? How could this be detected?
3. What chemicals could be put on pine cones or logs to get red and green flames in a fire?
4. What is the origin of fireworks? How do they work?

The Gas Laws and Everyday Life

Of the three common phases of matter, gases have held the greatest fascination for scientists. Unlike liquids and solids, the volume of a gas can be altered dramatically by changing its pressure or temperature. This is not just of interest to scientists. It is important in two very different areas of our lives—baking and scuba diving. They might seem to have nothing in common, but they both depend upon the fact that gases occupy a larger volume at a lower pressure.

When you bake a cake, a chemical reaction in the mixture produces carbon dioxide gas. This reaction makes all the tiny bubbles you see in a baked cake. If you bake a cake at a high altitude (such as Edmonton) and follow the sea-level recipe, the results will be very different. The same number of moles of carbon dioxide gas will be formed, but at the higher altitude the air pressure is lower. As a result, the carbon dioxide released in the cooking process will expand to a greater volume—more than the cake mixture can hold. The cake will expand like a balloon and then collapse. So, for successful baking at a high altitude, you need to use less baking soda in the mixture.

Why is scuba diving similar? As you dive deeper and deeper, the weight of the surrounding water puts more and more pressure on the air in your lungs. As you return to the surface, the pressure decreases and the air in your lungs will expand. If you held your breath while rising to the surface, your lungs could explode, just as the bubbles explode in the cake mix at high altitude.

RELATED EXPERIMENTS

Experiment 7 Boyle's Law and the Properties of Gases
Experiment 8 Charles' Law and the Properties of Gases

Experiment 7

Boyle's Law and the Properties of Gases

INTRODUCTION

Although we rarely see gases around us, they are extremely important substances. You may have seen how the volume of a gas can be affected by its pressure (for example, in your bicycle tires). This experiment illustrates the exact, quantitative relationship between the pressure and the volume of a gas. This relationship is called Boyle's Law.

PROBLEM

What quantitative effect does an increase in pressure have on the volume of a gas?

APPARATUS AND MATERIALS

Per pair of students:

1 barometer
1 60-mL syringe
1 pressure gauge

PROCEDURE

1. Measure and record the barometric pressure in the room.
2. Make sure the syringe is almost full of air. Attach the pressure gauge to the syringe.
3. Adjust the plunger in the syringe until the pressure gauge reads zero. Note the volume at which the gauge reads zero as shown in the diagram below.

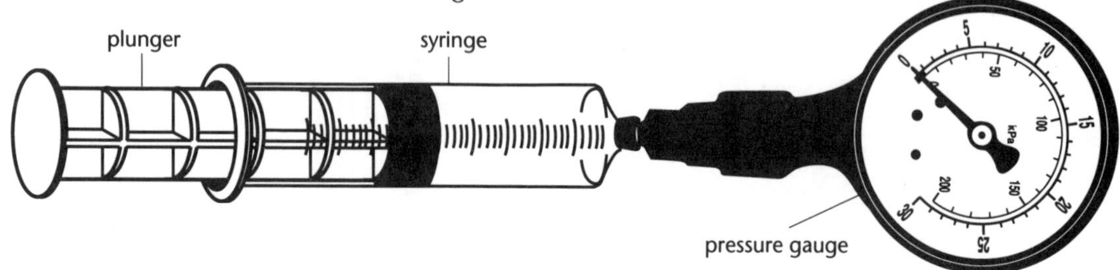

4. Push in the plunger to reduce the volume in the syringe by about half, so that the syringe plunger is at 30 mL on the volume scale. Hold the plunger steady and record in a table the pressure gauge reading and the volume of air in the syringe (30 mL).
5. Release the plunger slightly. Record in a table three more pressure values and corresponding volumes as you return the air in the syringe to the volume at which the pressure gauge read zero. The gauge and fittings also contain a small but significant volume of air. Your teacher will provide you with this volume. This should be added to each volume reading.
6. Repeat Steps 1 to 5 to make sure there is no leak in the apparatus. (You should get similar readings.)

Optional: Use a computer spreadsheet to record and analyse your data.

ANALYSIS

1. What pressure is the pressure gauge measuring?
2. Find the total pressure on the air in the syringe.
3. To how many significant figures did you measure the volume and pressure? Use this when completing the table of total pressures and total volumes, and round as necessary. Try to find some simple mathematical relationship between the pressure and volume. This is Boyle's Law.
4. Which set of data of pressure and volume most closely fits the above relationship?

EXTENSION

1. What pressure of air do you put in your bicycle tires?
2. (a) What does your tire look like when the gauge reads zero?
   (b) What is the true pressure in your tire then?
3. When a person's gas bill is calculated, barometric pressure is included (for example, in Stratford, Ontario, it is 0.982). What does this mean and why is it included?
4. If the total pressure on a gas doubles, what happens to the volume of the gas?

Experiment 8

Charles' Law and the Properties of Gases

INTRODUCTION

Temperature affects the volume of a gas. You probably know that gases expand when heated. This experiment illustrates the quantitative relationship between the temperature and the volume of a gas, known as Charles' Law.

PROBLEM

What quantitative effect does an increase in temperature have on the volume of a gas?

APPARATUS AND MATERIALS

Per pair of students:

1 10-mL syringe
1 stirrer
1 plug
1 250-mL beaker
1 thermometer
hot water
ice-cold water

PROCEDURE

1. Fill a plastic syringe about $\frac{2}{3}$ full of air and place a plug on the end to seal the air in.
2. Push down slightly on the plunger and release it, several times, to get an accurate reading of the volume. Record this value.
3. Measure the room temperature.

4. Fill a 250-mL beaker with ice-cold water. Place the syringe in the beaker so the trapped air in the syringe is under the water but the water is not over the syringe barrel, as shown in the diagram below.

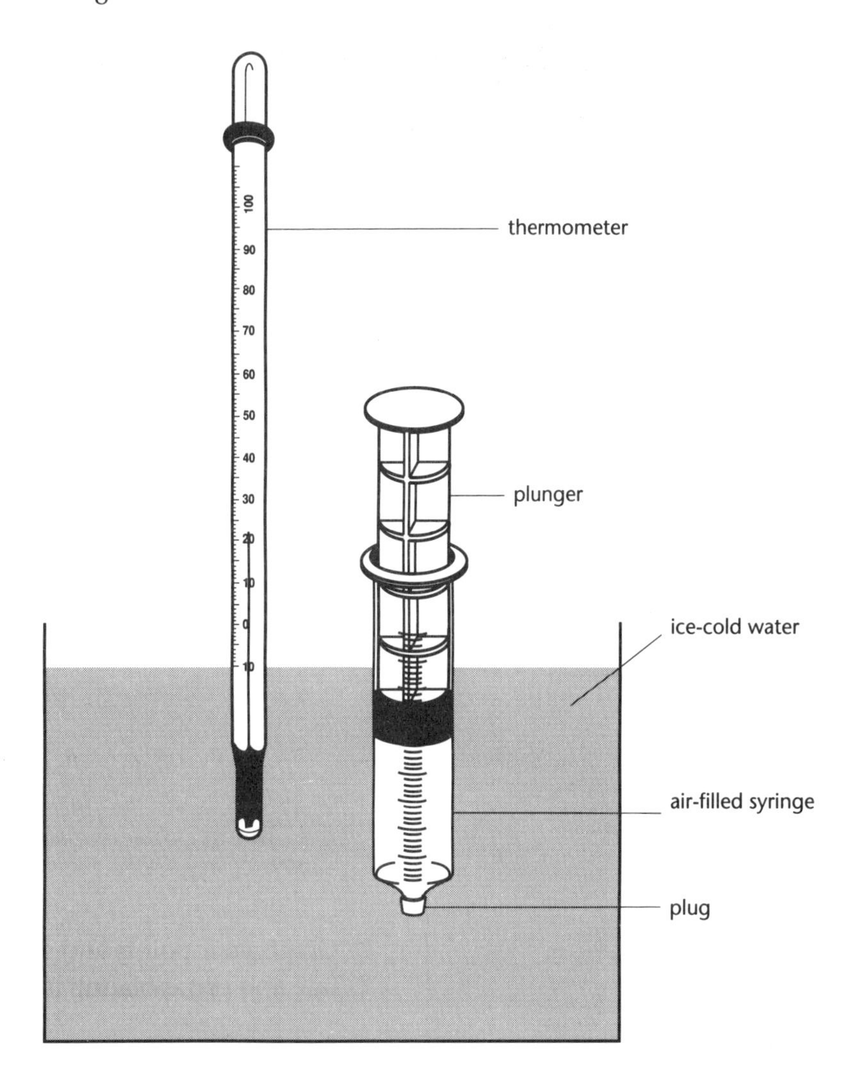

5. Stir the water well. Wait about one minute, then measure and record the temperature of the water. Push down on the plunger as in Step 2 to measure and record the volume of the air.

6. Remove the syringe and hold it carefully while your partner pours out about $\frac{1}{4}$ of the cold water and adds some warm water. Repeat Step 5.

7. Repeat Step 6 at several different water temperatures up to about 70°C (or the temperature of the hot water in your school). You should get a rise of about 8°C each time.

ANALYSIS

1. Graph volume V in mL (y-axis) against temperature T (x-axis) as instructed by your teacher. See the diagram below.

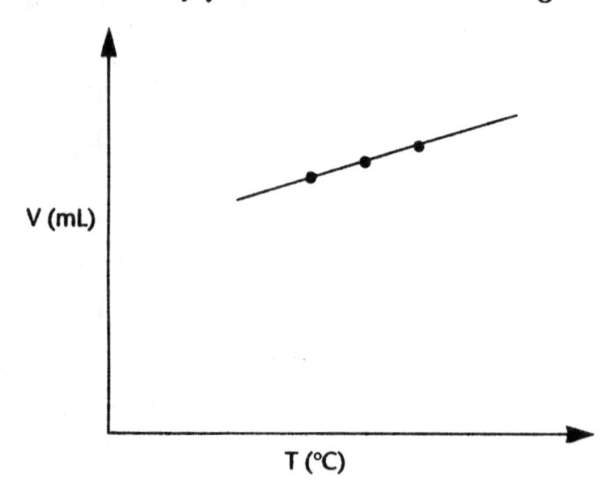

2. On another piece of graph paper, draw the axes as shown below.

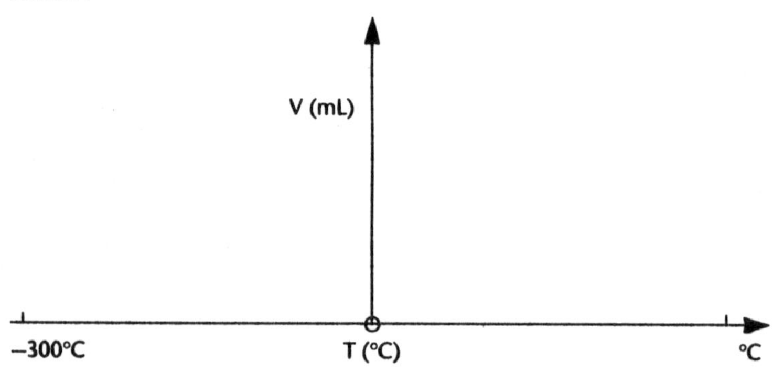

Graph your points and see your teacher before drawing in the line and extrapolating it.

3. Draw a third graph by placing the *V*-axis on the far left of the *T*-axis as shown below. Label the *T*-axis "total temperature".

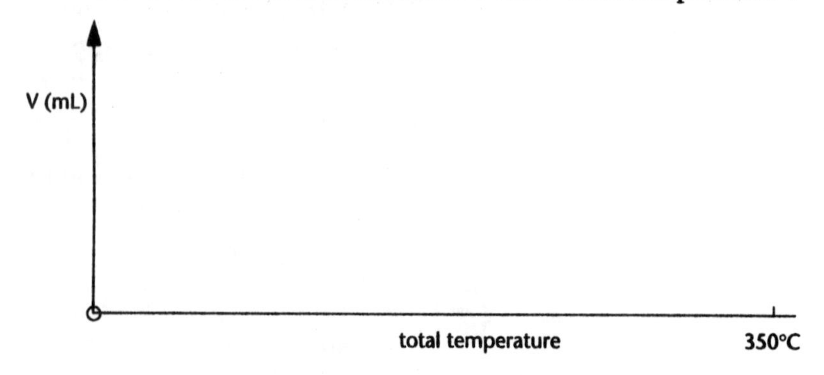

Place the third graph over the second graph so that the origin of the third graph is on top of the *T*-intercept on the second graph, and the temperature axes are superimposed. Mark all points on the third graph that are on the second graph. Draw a line through these points.

4. Using this graph, make a table of the volume values at "total temperatures" of 50, 100, 200, and 300. How are the "total temperature" and volume related?

5. (a) Qualitatively, as the temperature rises what happens to the volume?
   (b) Quantitatively, how are volume and temperature related? This is Charles' Law.

6. What is another name for the "total temperature"?

7. If the "total temperature" doubles, what happens to the volume?

8. (a) What was your value of the *T*-intercept on the Celsius temperature scale?
   (b) At the *T*-intercept, what is the volume of the gas? Is this possible? What would happen to the gas before you reached this temperature?

EXTENSION

Using what you have learned about Boyle's Law and Charles' Law, describe an efficient way of storing a large amount of gas.

Experiment 9

Molar Volume of a Gas

INTRODUCTION

The volume occupied by a sample of gas depends on the mass of the gas, its pressure, and its temperature. To compare the volumes of different gases we need a standard; in chemistry the standard is 1 mol of a substance.

PROBLEM

What is the volume of 1 mol of air?

APPARATUS AND MATERIALS

Per pair of students:

1 60-mL syringe
1 plug
1 stick
1 balance

PROCEDURE

1. Put the plug on the end of the syringe, after the plunger has been fully pushed into the syringe. With the plug on, pull back the plunger to the point where you can insert the stick to keep the plunger at a fixed position as shown in the diagram below. Measure and record the mass of the empty syringe, stick, and plug.

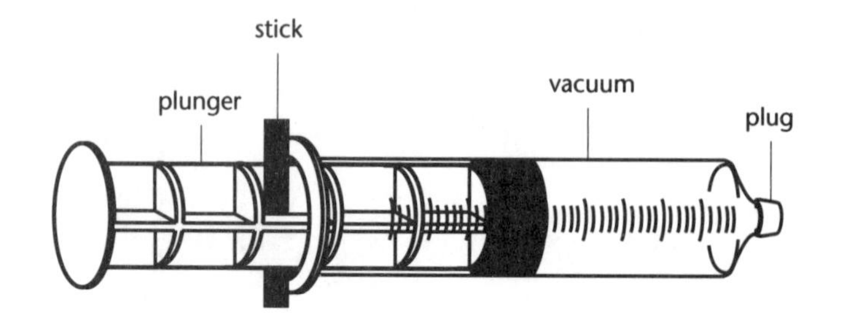

2. With the stick still in place, remove the plug and let the syringe fill with air, making sure that the plunger is in exactly the same position as in Step 1. Replace the plug.

3. Measure and record the mass of the syringe containing air.

4. Measure and record the atmospheric pressure and room temperature.

Optional: Use a computer spreadsheet to record and analyse your data.

ANALYSIS

Use the ideal gas law, $PV = nRT$, to do your calculations.

1. What is the mass of air in the syringe?

2. (a) The mass of 1 mol of air is 28.9 g. Find the volume of 1 mol of air at atmospheric pressure and room temperature.
   (b) Convert this volume to standard temperature and pressure.
   (c) How many significant figures should your answer have?

EXTENSION

1. Compare your value for the volume of 1 mol of air to the volume of 1 mol of each of the following gases at standard temperature and pressure.

Gas	Density at STP
hydrogen	0.0899 g/L
carbon dioxide	1.98 g/L
methane	0.717 g/L
oxygen	1.43 g/L

2. How do the molar volumes of different gases compare?

3. Using this apparatus and your results, how could the molar mass of another gas be found?

Experiment 10

Molar Mass of a Gas

INTRODUCTION

In an earlier experiment you should have found that all gases, at standard temperature and pressure (STP), occupy the same volume. Also, Avogadro's hypothesis states that equal volumes of two gases, at the same temperature and pressure, contain the same number of molecules. Thus, by comparing a gas whose molar mass is unknown to a gas whose molar mass is known, you can find the unknown molar mass. When you know the molar mass, you can find the molecular formula.

PROBLEM

What is the molar mass and molecular formula of butane?

APPARATUS AND MATERIALS

Per pair of students:

1 60-mL syringe
1 plug
1 balance
1 stick
1 can of butane

SAFETY

Butane is flammable. Make sure there are no open flames in the laboratory.

PROCEDURE

1. Put the plug on the syringe after the plunger has been pushed all the way in. With the plug on, pull back the plunger until it can be secured with a stick as shown in the diagram for Experiment 9. Measure and record the mass of the empty syringe, plug, and stick.

2. With the stick still in place, remove the plug and let the syringe fill with air, making sure the plunger is in the same position as in Step 1. Replace the plug. Record the mass of the air-filled syringe.

4. Remove the plug from the syringe. Push the air out of the syringe and refill the syringe with butane. Let the butane pressure push the plunger back until you can place the stick into the plunger. Disconnect the butane and replace the plug on the syringe. This volume of butane should equal the volume of the air you measured previously.

5. Measure and record the mass of the syringe and butane

6. Dispose of the butane as instructed by your teacher.

Optional: Use a computer spreadsheet to record and analyse your data.

ANALYSIS

1. Calculate the masses of the equal volumes of air and butane.
2. Air has a molar mass of 28.9 g. Find the molar mass of butane.
3. Analysis of butane shows that it contains 82.8% carbon and 17.2% hydrogen. Find the simplest formula of butane.
4. Find the molecular formula of butane.

EXTENSION

1. Using the mass listed on the can of butane, calculate the amount (in moles) of butane in the can.
2. Why would propane not be used in lighters?
3. Draw a structural formula of butane.
4. How could the molar mass of a liquid or solid be found? If you have a computer and an appropriate program, you could use a simulation to do this.

Chemical Reactions on an Industrial Scale

Most of the experiments in this laboratory manual concern chemical reactions. You perform a reaction to make a product and then you safely dispose of the small amounts of the products produced. But doing chemical reactions on an industrial scale is not that easy. For example, the steel industry turns brown, powdery iron ore (iron(III) oxide containing traces of other elements) into shining steel. In the process, iron(III) oxide reacts with carbon at high temperatures to give iron metal and a mixture of carbon monoxide and carbon dioxide.

The naturally occurring iron ore contains many impurities, such as sand. Calcium carbonate is added to the reaction mixture. The calcium carbonate reacts with the sand to form calcium silicate (slag) that can be drained off leaving pure steel. Unfortunately, slag has little use except as fill for road making. In addition, million-tonne quantities of carbon dioxide and carbon monoxide are produced in the reaction. These gases have to be disposed of without polluting the air. Finally, the steel produced is extremely hot and so millions of litres of water are needed to cool the metal (though in modern plants, this water is recycled as much as possible). Thus, chemistry on an industrial scale involves much more thought than chemistry on a laboratory scale.

As you perform the microscale reactions in these experiments, think how they could be adapted to produce products in the thousands or millions of tonnes.

RELATED EXPERIMENTS

Experiment 11 Single Displacement Reactions
Experiment 12 Double Displacement Reactions

Experiment 11

Single Displacement Reactions

INTRODUCTION

In a single displacement reaction, an element reacts with a compound to produce a different element. An example is the reaction of zinc metal with nickel(II) chloride.

$$Zn(s) + NiCl_2(aq) \longrightarrow ZnCl_2(aq) + Ni(s)$$

In this experiment, you will combine metals and metal ion solutions to find which metals are the most reactive.

PROBLEM

Which metals will displace other metal ions?

APPARATUS AND MATERIALS

Per pair of students:

1 96-well plate
1 1-mL micro-tip pipette
4 strips copper metal
4 strips lead metal
4 strips zinc metal
1 piece of emery paper
dropper bottles of:
  silver nitrate solution
  zinc sulfate solution
  lead(II) nitrate solution
  copper(II) sulfate solution

SAFETY

Lead and lead salts are toxic and can cause birth defects if ingested by pregnant women. However, in the dilute solutions used here, they present little risk. As an extra precaution, wash your hands thoroughly after the experiment. Silver nitrate can stain skin and clothing.

PROCEDURE

1. Rub the copper and zinc strips gently with the emery cloth to clean the surfaces. Do not rub the lead strip; your teacher has cleaned it. Make sure you know which strip is which metal.
2. Place 2 drops of each of the following solutions in each of the wells as indicated below.

Solution	Wells
silver nitrate	A1 to A3
zinc sulfate	B1 to B3
lead(II) nitrate	C1 to C3
copper(II) sulfate	D1 to D3

3. Place a clean strip of metal in each well as indicated below.

Metal	Wells
copper	A1 to D1
zinc	A2 to D2
lead	A3 to D3

4. Leave the strips in the solutions for about 30 s. While you are waiting, draw up a table with the metals across the top and metal solutions down the side. Use this table to record your observations.
5. Remove the metal strips, one at a time, and observe whether there is any change in the metal surface or in the colour of the solution. Record your observations in the table.

6. Dispose of the lead-containing solutions as instructed by your teacher.

ANALYSIS

1. (a) According to your observations, which metal reacted with the most solutions?
   (b) Which metal ion reacted with all the metals?
2. (a) Of the metals and metal ions that you studied in this experiment, which metal reacted with the greatest number of solutions?
   (b) Which metal reacted with the fewest solutions? Explain your reasons.
3. In the cases where a reaction happened, write a balanced chemical equation to represent each reaction.

EXTENSION

1. Obtain from your teacher other metals and other metal ion solutions that you can use to extend your table of results.
2. Hydrogen ions can undergo single displacement reactions to give hydrogen gas. For example, magnesium metal reacts with hydrochloric acid.

$$Mg(s) + 2HCl(aq) \longrightarrow MgCl_2(aq) + H_2(g)$$

Repeat the above experiment and attempt to react the three metals with dilute hydrochloric acid. Which metals react with acid and which do not? Be careful when using hydrochloric acid. It is corrosive.

Experiment 12

Double Displacement Reactions

INTRODUCTION

In a double displacement reaction, two compounds react such that the cations and anions switch partners. An example is the reaction of hydrochloric acid with sodium hydroxide solution.

$$HCl(aq) + NaOH(aq) \longrightarrow NaCl(aq) + H_2O(l)$$

As you can see from the above reaction, one reason for a double displacement reaction occurring is the formation of water. However, there are two more driving forces, and you will identify these as you conduct the experiment.

PROBLEM

What causes a double displacement reaction to occur?

APPARATUS AND MATERIALS

Per pair of students:

1 96-well plate
1 toothpick
paper towel
dropper bottles of:
- calcium chloride solution
- copper(II) sulfate solution
- cobalt(II) nitrate solution
- sodium carbonate solution
- dilute hydrochloric acid solution

SAFETY

Hydrochloric acid is corrosive. However, in the dilute solution used here, it presents little risk. Wash any spills off skin and clothing.

PROCEDURE

1. Place 1 drop of each solution in a well as indicated below.

Solution	Well
calcium chloride	A1
copper(II) sulfate	A2
cobalt(II) nitrate	A3
hydrochloric acid	A4

2. Add 1 drop of the sodium carbonate solution to each well. Stir the contents of each well with a toothpick. Note your observations. Wipe the toothpick with a paper towel before placing it in the next well.

ANALYSIS

1. Apart from the formation of water, what are the two other driving forces for a double displacement reaction?
2. Write a balanced chemical equation for each precipitation reaction.
3. (a) Write a balanced chemical equation for the reaction producing a gas.
   (b) What gas was formed? Devise a method for testing your hypothesis.

EXTENSION

Your teacher will provide you with other salts. Using a table of solubilities, determine what combinations of soluble salts would also give precipitates. To confirm your predictions, repeat this experiment using these salts.

The Periodic Table

The Periodic Table is the most important tool used in studying chemistry. Without this ingenious chart, you would have to learn the chemical characteristics of over one hundred different elements. When you look at the Periodic Table, you can see many patterns and trends in the different elements and groups of elements.

These trends in the Periodic Table are not just theoretical. They can have very real consequences. Consider the Chernobyl nuclear power plant in the former Soviet Union. In 1986, due to an accident, the power plant released a huge cloud of radioactive material. The cloud contained a radioactive isotope of the element strontium. Strontium is particularly dangerous because it is in the same group as calcium in the Periodic Table. Because strontium has the same properties as calcium, the body uses strontium the same way it normally would use calcium. So if someone ate food contaminated with strontium, the body would deposit it in the bones as it normally would deposit calcium. The radiation from the strontium would damage the bone marrow, leading to leukemia and possible death. By knowing the patterns in the Periodic Table, scientists could predict the effect the radioactive cloud would have on plants, animals, and humans. In this case, people living in the path of the radioactive cloud were advised to take massive doses of calcium supplements to minimize the possibility of their bodies absorbing the radioactive strontium.

RELATED EXPERIMENT

Experiment 13 Patterns in the Properties of Compounds

Experiment 13

Patterns in the Properties of Compounds

INTRODUCTION

The Periodic Table is one of the most useful summaries of chemical information. In this experiment, you will investigate some properties of Group II ions in the Periodic Table, and look for any patterns. You will use these patterns to identify an unknown element.

PROBLEM

What are the patterns in the properties of the compounds of Group II ions?

APPARATUS AND MATERIALS

Per pair of students:

1 96-well plate
12 1-mL micro-tip pipettes
0.10 mol/L solutions of:
- magnesium nitrate
- calcium nitrate
- strontium nitrate
- barium nitrate

0.20 mol/L solutions of
- sodium chloride
- potassium chromate
- sodium sulfate
- sodium acetate (ethanoate)
- sodium bromide
- sodium hydroxide
- sodium carbonate

2 samples of an unidentified solution

SAFETY

Barium ions are toxic. The chromate ion is a suspected carcinogen. Sodium hydroxide is corrosive. However, in the concentrations used here, they present little risk. As an extra precaution, wash your hands thoroughly after the experiment.

PRELAB ASSIGNMENT

Make two tables on two separate pages with the same headings. Down the left side, list the four Group II ions by name and formula. Across the top, list the seven anions (acetate, bromide, carbonate, chloride, chromate, hydroxide, sulfate) and their formulas.

In the first table, record your observations.

PROCEDURE

1. In a 96-well plate, use micro-tip pipettes to place 3 drops of each Group II ion solution in your plate in the same order as in your table. For example, place the magnesium nitrate solution in wells A1 to A7. Place the calcium nitrate solution in wells B1 to B7. As each solution is added, note it on your table.
2. Place 3 drops of each anion solution in 4 wells, vertically. For example, place the acetate ion solution in wells A1 to D1. Place the bromide solution in wells A2 to D2, and so on. Remember to check off each solution as it is added.
3. Hold your well plate at right angles to a light source so that you can observe carefully any reactions that occur (precipitates that form). Note relative amounts and other characteristics.

ANALYSIS

1. In the second table, write the names of the possible compounds formed, and their formulas.
2. (a) Which anions did not form a precipitate (did not react)?
   (b) Which ions did form a precipitate?
3. You should now know what anions to use to identify a solution that contains one of the Group II ions. Your teacher will give you two samples of an unidentified solution. (Use 3 drops each.)
4. Conduct two tests to identify the unknown solution.
5. What is the unknown solution? Give reasons for your answer.

EXTENSION

A more elaborate method of identifying substances is called *qualitative analysis*. This can be done for either cations or anions. Find out how this is done. You may wish to consult your teacher or conduct research.

Experiment 14

Stoichiometry of a Reaction Producing a Solid

INTRODUCTION

Chemical reactions are one of the most important aspects of chemistry. Stoichiometry is the study of how much of one reagent reacts with another. In this experiment, you will compare the mass of iron consumed in a reaction to the mass of copper produced.

PROBLEM

How does the mass of a product relate to the mass of a reactant?

APPARATUS AND MATERIALS

Per pair of students:

2 30-mL or 50-mL beakers
1 20-mL graduated cylinder
filter apparatus (see the diagram on the next page)
1 jumbo pipette
small filter paper (about 4 cm in diameter)
1 balance
1 toothpick
1 paper clip
powdered copper(II) sulfate pentahydrate
iron powder
distilled water

PRELAB ASSIGNMENT

Write a balanced chemical equation for the reaction between copper(II) sulfate pentahydrate solution and iron metal. (Hint: One of the products is iron(II) sulfate.)

PROCEDURE

1. Measure and record the mass of the beaker. Add powdered copper(II) sulfate pentahydrate. Your teacher will tell you how much to add. Measure and record the mass of the beaker and its contents. Calculate the precise mass of copper(II) sulfate.
2. Add between 15 mL and 20 mL of distilled water. Stir until the powder is completely dissolved.
3. Weigh your assigned mass of iron powder. Record its mass.
4. Add the iron powder to the copper(II) sulfate solution prepared in Step 2. Stir for about 5 min. Make sure that you break up any solid lumps on the bottom of the beaker.
5. In between stirrings, prepare your filter apparatus. Find the mass of a filter paper. Place the eraser end of a pencil in the centre of the paper and bend the paper up around the body of the pencil. The paper should now be in the shape of a tube with one sealed end. Place the folded paper into the filter apparatus. Suspend the filter apparatus over the inside of the other beaker using a twisted paper clip as shown in the diagram below.

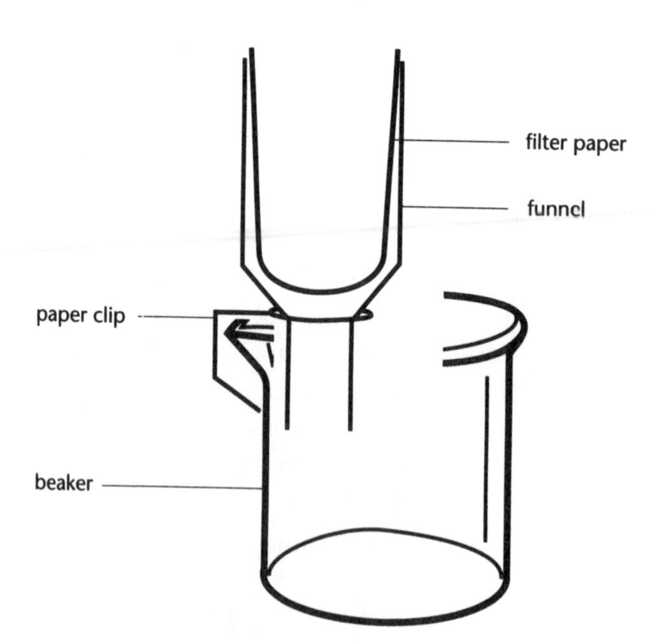

6. Let the copper metal settle and pour off as much of the solution as you can. Then use the jumbo pipette to transfer the solid and the remaining liquid to the filter.
7. Squirt some distilled water into the beaker and draw up any remaining solid with the jumbo pipette. Transfer it to the filter paper.
8. Leave the paper containing the copper to dry overnight and then find the mass of the copper and filter paper. Calculate the mass of copper produced.

ANALYSIS

1. Collect data from other students. Plot the mass of iron (y-axis) against the mass of copper (x-axis). Include the point (0,0) (that is, 0 g of iron would produce 0 g of copper) on your graph. Draw the line of best fit. What does the slope of the line represent?

2. Calculate the ratio of molar mass of iron to molar mass of copper. How does this compare with your results?

Optional: Use a computer graphing program to analyse your data.

ENVIRONMENTAL APPLICATION

This reaction has industrial importance. In the copper mining industry (and at the Canadian Mint), scrap iron is used to remove copper ions from waste waters, replacing them with the relatively harmless iron ions. (Copper ions are toxic to aquatic organisms.) Investigate where in the country copper mining and refining is done.

Experiment 15

Stoichiometry of a Reaction Producing a Gas

INTRODUCTION

The Law of Constant Composition (Definite Proportions) states that for any compound that is made or is decomposed, there is a precise amount of each product that is produced. In this experiment, you will react sodium bicarbonate with sulfuric acid to form a salt, carbon dioxide, and water. You will then use the Law of Constant Composition to calculate the amount of carbon dioxide produced.

PROBLEM

How many moles of carbon dioxide are produced from 1 mol of sodium bicarbonate?

APPARATUS AND MATERIALS

Per pair of students:

1 24-well plate
1 balance
1 1-mL micro-tip pipette
1 spatula
2.00 mol/L solution of sulfuric acid
sodium bicarbonate

SAFETY

Sulfuric acid is corrosive. Avoid contact with skin and clothing. Flush any contacted area with running water.

PROCEDURE

1. Fill a pipette with a 2.00 mol/L sulfuric acid solution.
2. Place a 24-well plate on the balance and place the acid-filled pipette in well C3 with the bulb end down.
3. Zero the balance and use a spatula to add less than 0.3 g of sodium bicarbonate to well B3, as shown in the diagram below.

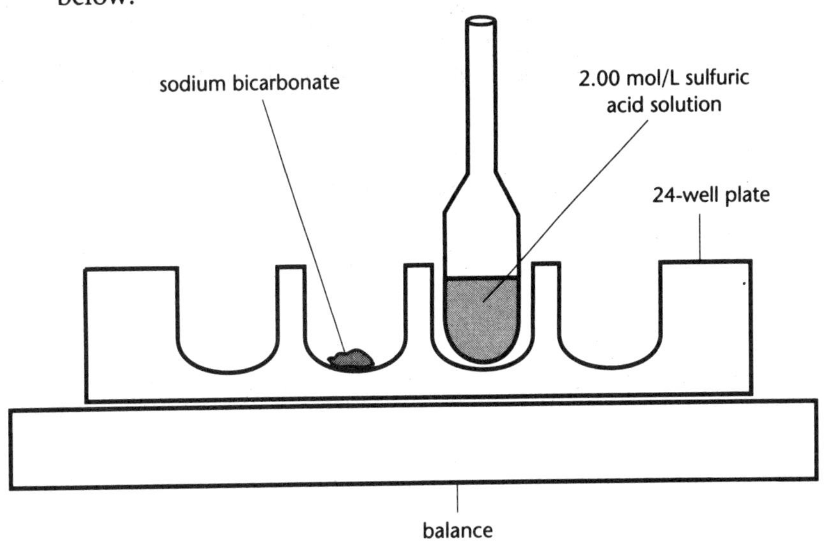

4. Record the mass of sodium bicarbonate used.
5. Zero the balance.
6. Remove the pipette from well C3 and add the sulfuric acid to the sodium bicarbonate, 1 drop at a time. After each drop, allow the reaction to subside.
7. Continue adding sulfuric acid, 1 drop at a time, until all the sodium bicarbonate has reacted and there is no further effervescence.
8. Replace the pipette back in the plate and measure the mass. Calculate the loss in mass.

Optional: Use a computer spreadsheet to record and analyse your data.

ANALYSIS

1. Use the mass of the sodium bicarbonate to find the number of moles of sodium bicarbonate.
2. What was the mass of carbon dioxide produced? Convert this mass from grams to moles.
3. For 1 mol of sodium bicarbonate, how many moles of carbon dioxide are produced? In most chemical equations we use whole numbers. What is your value for the number of moles of carbon dioxide produced, to the nearest whole number?

EXTENSION

1. (a) Write an equation for the reaction of sulfuric acid with sodium bicarbonate.
   (b) Use the value for the number of moles of carbon dioxide produced to balance the equation.
   (c) What volume of carbon dioxide was produced by the reaction?
2. (a) Why is sodium bicarbonate used in baking?
   (b) What else must be present as an ingredient to help the sodium bicarbonate do what it is supposed to?
   (c) For the dough to rise, what consistency must it have to trap the gas?
3. Find a recipe at home that uses sodium bicarbonate (baking soda) and calculate the amount of carbon dioxide that would be produced. If the recipe is for a dozen items, calculate the volume of carbon dioxide produced for each item.
4. In 1984, a man in the U.S. sued a baking soda manufacturer because his stomach burst after taking 1 teaspoon of baking soda. Calculate what volume of gas, at stomach conditions, was produced.

Experiment 16

Production of Hydrogen and Oxygen

INTRODUCTION

Hydrogen and oxygen are two very important gases in our world. Among other things, oxygen is used to oxidize rocket fuels and to process metals, such as iron, in furnaces. Hydrogen is used to upgrade oil, make margarine, and it may become the fuel of the future. In this experiment, you will make both gases and then see how they react.

PROBLEM

In what mole ratio do hydrogen and oxygen react?

APPARATUS AND MATERIALS

Per pair of students:

1 24-well plate
1 stopper and delivery tube
5 gas-collecting pipettes
1 burner
1 1-cm piece of magnesium ribbon
1.0 mol/L solution of hydrochloric acid
pyrolusite rock
hydrogen peroxide solution

SAFETY

Hydrogen peroxide is very corrosive. Handle with care. Hydrochloric acid is also corrosive. Avoid contact with skin and clothing. Flush any contacted area thoroughly with running water.

PRELAB ASSIGNMENT

1. Hydrogen is produced by the single displacement of magnesium with hydrochloric acid. Write a balanced equation for this reaction.
2. Write the balanced equation for the decomposition of hydrogen peroxide.

PROCEDURE

PART I—HYDROGEN PRODUCTION

1. Fill the five gas-collecting pipettes with tap water and place them into wells A1 to A5. If the graduations are not already marked on the pipettes, mark the graduations on the pipettes as shown below.

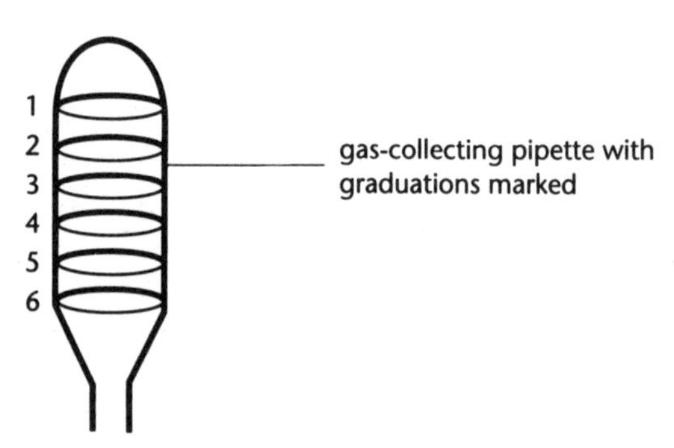

2. Fill well F1 about $\frac{1}{3}$ full of hydrochloric acid and then add a 1-cm piece of magnesium ribbon. Immediately place the stopper and delivery tube on top of the well.
3. Wait about 15 s for any air to be displaced from the well. Put the first gas-collecting pipette on top of the delivery tube as shown in the diagram below.

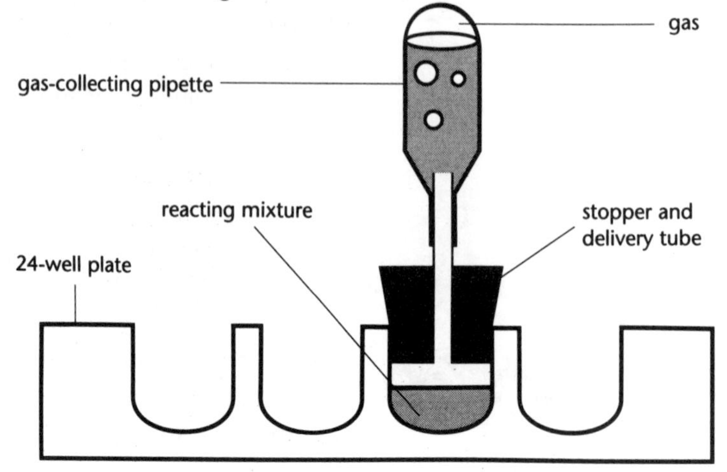

4. Collect 5 parts of hydrogen and then replace the pipette back in well A1. The remaining water in the pipette will act as a plug for the collected gas.

5. Repeat Step 4 using the second gas-collecting pipette, but collect only 4 parts of hydrogen. Place the pipette in well A2.

8. Continue to collect 3 parts, 2 parts, and 1 part of hydrogen in the remaining three pipettes. Place these pipettes in wells A3 to A5, respectively.

PART II—OXYGEN PRODUCTION

9. Fill well F6 about $\frac{1}{3}$ full of hydrogen peroxide solution.

10. Add a piece of pyrolusite rock.

11. Place the stopper and delivery tube into this well.

12. Wait 15 s for the oxygen gas to displace the air.

PART III—HYDROGEN–OXYGEN REACTION

13. Place the gas-collecting pipette containing hydrogen from well A1 on top of the delivery tube. Fill the pipette *almost* full, leaving a small drop of water in the opening to keep the gases in. Replace the pipette in well A1.

14. Fill the remaining pipettes with oxygen in a similar manner. Replace each pipette in its well. You should end up with the five pipettes filled with different mixtures of hydrogen and oxygen.

15. Make a table and record how much of each gas is in each pipette. Include a column to record how loud the reaction is (none, weak, moderate, loud, very loud).

16. Light a burner. Hold the first pipette horizontally with its opening about 2 cm from the heat source as shown in the diagram below. Quickly squeeze the pipette and note the loudness of the reaction.

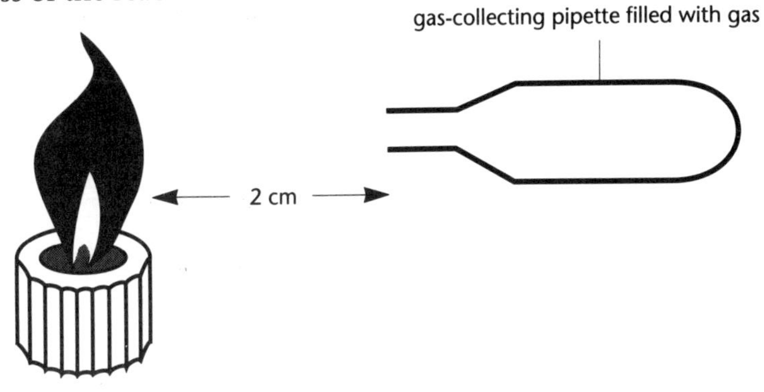

17. Repeat Step 4 for the remaining four pipettes.

ANALYSIS

1. Draw a bar graph with the loudness on the *y*-axis, and the ratio of hydrogen volume to oxygen volume on the *x*-axis.
2. What ratio of hydrogen to oxygen produced the loudest sound?
3. From Avogadro's hypothesis, what do we know about the number of molecules in equal volumes of two gases?
4. Write a balanced equation for the reaction of hydrogen and oxygen.

EXTENSION

1. Hydrogen peroxide is a bleach and a disinfectant.
   (a) From the Analysis, what substance produced from hydrogen peroxide causes the bleaching effect?
   (b) What else is produced? Think of some reasons why hydrogen peroxide is being used more frequently to bleach paper and clothes and, in the future, might be used to clean our water supply.
2. (a) What sound would pure hydrogen produce when squirted into a flame?
   (b) What sound would pure oxygen produce when squirted into a flame?
3. Read about the use hydrogen as a fuel. What advantages does it have over gasoline or propane? What disadvantages need to be overcome to be able to use hydrogen as a fuel?

Efficiency in Industrial Processes

The chemical industry is extremely competitive. The prices of products that the chemical industry produces are determined by market forces. If the cost of producing a chemical at a plant is greater than the current market price, then there is little option but to close down that plant. One way to decrease costs is to increase the percentage yield of a reaction. Researchers constantly look for ways to optimize production by altering the reaction conditions so that the maximum possible yield of products is obtained. For example ammonia, which is an important crop fertilizer, is produced in microscopic yields at atmospheric pressure. By increasing the pressure to about five hundred times that of atmospheric pressure, the percentage yield can be increased to about 30%.

Another way to save costs is to find uses for by-products. Sulfur dioxide is a pollutant generated during the production of several metals such as nickel, lead, and zinc. Instead of releasing this gas—a major contributor to acid rain—into the air, it is used to synthesize sulfuric acid. Thus, a harmful waste gas becomes an industrially important chemical that can be sold.

RELATED EXPERIMENT

Experiment 17 Percentage Yield and Purity

Experiment 17

Percentage Yield and Purity

INTRODUCTION

Many chemical reactions do not always form the products we expect nor the amounts we predict. One way to determine the efficiency of the reaction is to calculate the percentage yield. Percentage yield can also be used to check the purity of consumer products. For example, Tums is an antacid used to neutralize "excess stomach acid." The active ingredient is calcium carbonate which reacts with stomach acid, hydrochloric acid, to form calcium chloride, carbon dioxide, and water.

In this experiment, you will calculate how much calcium carbonate a Tums tablet contains.

PROBLEM

Is Tums pure calcium carbonate?

APPARATUS AND MATERIALS

Per pair of students:

1 24-well plate
1 5-mL micro-tip pipette
1 balance
1 spatula
Part of 1 crushed Tums tablet
4.00 mol/L solution of hydrochloric acid

SAFETY

Hydrochloric acid is very corrosive. Handle with care. Avoid contact with skin and clothing. Flush any contacted area thoroughly with running water.

PROCEDURE

1. Fill a 5-mL micro-tip pipette with a 4.00 mol/L solution of hydrochloric acid.
2. Place a 24-well plate on the balance and place the acid-filled pipette in well C3, with the bulb end down.

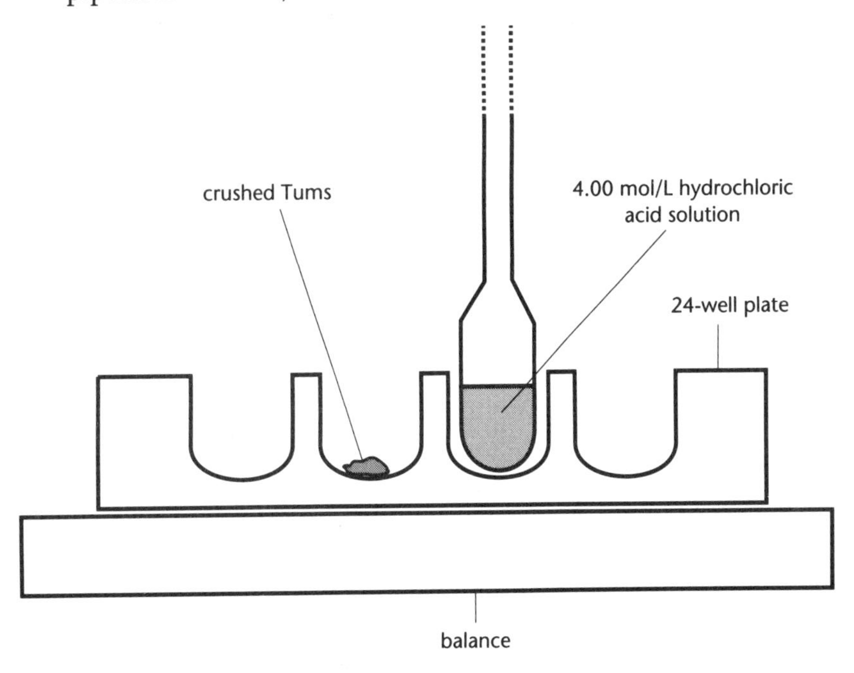

3. Zero the balance and use a spatula to add about 0.15 g of crushed Tums to well B3.
4. Record the mass of Tums used.
5. Remove the pipette from well C3. Carefully add the acid, 1 drop at a time, to the crushed Tums. As the reaction subsides, add more drops of acid until there is no more effervescence.
6. Replace the pipette in well C3 and record the mass again.

Optional: Use a computer spreadsheet to record and analyse your data.

ANALYSIS

1. Write a balanced equation for the reaction.
2. What mass of carbon dioxide should be produced if the Tums were pure calcium carbonate? This is the theoretical yield.
3. How much carbon dioxide was actually produced?
4. Calculate the percentage yield.

EXTENSION

1. Give two reasons for the low percentage yield.
2. Why was the Tums tablet crushed?
3. Suggest two reasons why you should swallow a Tums tablet whole and not chew it.

Experiment 18

Classifying Ionic Solutions

INTRODUCTION

Certain substances are usually soluble in water. In this experiment you will be using several different chemicals that are soluble in water and classifying them into a few small groups. This was how early scientists organized and classified materials.

PROBLEM

What physical property can be used to classify solutions?

APPARATUS AND MATERIALS

Per pair of students:

1 96-well plate
1 1-mL micro-tip pipette
9 toothpicks
9 watch glasses, each containing one of these:
copper(II) sulfate, cobalt(II) chloride, sodium chloride, potassium hydroxide, copper(II) nitrate, sodium nitrate, sodium hydroxide, cobalt(II) nitrate, sodium phosphate
dropper bottles containing 0.100 mol/L solutions of:
hydrogen chloride
hydrogen sulfate
hydrogen phosphate
distilled water

SAFETY

Solid sodium hydroxide is very corrosive. Handle with care. Potassium hydroxide, hydrogen chloride, hydrogen sulfate, and hydrogen phosphate are also corrosive. Avoid contact with skin and clothing. Flush any contacted area with running water.

PRELAB ASSIGNMENT

1. Draw up a table with the following headings:

Chemical Compound	Ions	Formula	Observations

2. Fill in the 12 chemical compounds listed above, their ions, and their chemical formulas.

PROCEDURE

1. In the 96-well plate, fill wells A1 to A12 about $\frac{1}{3}$ full with distilled water using a pipette.
2. Wet a toothpick by dipping it in well A1. Insert this toothpick into copper(II) sulfate so that some of it sticks to the toothpick. Stir the water in well A1 with the toothpick to make a solution. Repeat this procedure if necessary to form a concentrated solution. Remove the toothpick. Record your observations.
3. If a compound is already in solution, observe and record its appearance.
4. Repeat Steps 2 or 3 for the remaining 11 compounds.
5. Compare your solutions and consider how to classify them.

ANALYSIS

1. Into what three categories could you put these solutions?
2. What particles (for example, atoms, ions, or molecules) does each compound produce in solution?
3. For the category that contains the smallest number of solutions, can you see a pattern in the contents of the solutions? Hint: From your table, look at the ions present.
4. Can you see another pattern in the next largest category of solutions? What is this pattern?
5. What is the largest category of solutions?
6. Can you suggest two or three tests to sort these solutions further?

EXTENSION

1. (a) What ion, present in some natural water (as opposed to bottled mineral water), helps prevent tooth decay?
   (b) Would it matter if this ion was present in a sodium compound or a calcium compound or a tin compound in toothpaste? Explain using your answer to the Problem.

2. What ions cause the hardness in water? What effect do these ions have on soaps and detergents?

Experiment 19

Conductivity of Solutions

INTRODUCTION

All matter is made up of tiny particles called atoms. An element is made up of one kind of atom. A compound consists of two or more different atoms bonded together. When a new compound is formed, it is different from the original elements or any compounds from which it was formed. One property that can illustrate this is the solubility of the compound and, in particular, its solubility in water.

Although all compounds are soluble in water to a lesser or greater extent, not all compounds dissociate in water to form freely moving ions. Some compounds may ionize in water, while other compounds neither dissociate nor ionize.

One way to assess the dissociation/ionization tendency of a compound is to check the conductivity of its aqueous solution. A compound whose aqueous solution conducts electricity is called an electrolyte, whereas a compound that forms a non-conducting solution is called a non-electrolyte.

PROBLEM

Which solutions or liquids conduct electricity?

APPARATUS AND MATERIALS

Per pair of students:

1 96-well plate
11 1-mL micro-tip pipettes
1 conductivity tester
0.100 mol/L solutions of:
- sodium chloride
- hydrochloric acid
- potassium nitrate
- sucrose

sodium hydroxide
barium hydroxide
sulfuric acid

ethanol
pure acetic (ethanoic) acid
tap water
distilled water

SAFETY

Pure acetic acid is very corrosive. Handle with care. Sodium hydroxide, sulfuric acid, hydrochloric acid, and barium hydroxide are also corrosive. Avoid contact with skin and clothing. Flush any contacted area thoroughly with running water for several minutes. Barium ions are toxic. However, in the small amounts used here, they present little risk. As an extra precaution, wash your hands thoroughly after the experiment. Ethanol is flammable. Make sure there are no open flames in the laboratory.

PROCEDURE

1. In the 96-well plate, place 6 drops of the solutions of sodium chloride, potassium nitrate, sodium hydroxide, hydrochloric acid, sucrose, barium hydroxide, and sulfuric acid in wells A1 to A7, respectively using a different pipette for each solution.
2. Place 6 drops of tap water in well A8.
3. Place 6 drops of ethanol in well A9.
4. Place 6 drops of distilled water in wells A10 and A11. Fill well H1 with distilled water.

5. Place 6 drops of pure acetic acid in well A12.
6. Place the conductivity tester in well A1 and record the intensity of the light. Rinse the tester in well H1 and dry it before each test.
7. Repeat Step 6 for each of the other liquids or solutions in wells A2 to A10 and A12.
8. Fill a pipette with barium hydroxide solution from the dropper bottle.
9. Place the conductivity tester in the sulfuric acid solution in well A7. Record the light intensity.
10. Add the barium hydroxide solution drop by drop to well A7 while stirring with the tester. Observe and record what happens to the light intensity and to the solution.

11. Repeat Steps 8 to 10, replacing the barium hydroxide with ethanol, adding it to the distilled water in well A10.
12. Repeat Steps 8 to 10 using acetic acid and adding it to the distilled water in well A11.

ANALYSIS

1. Which solutions conducted well? Why?
2. Which solutions were non-conducting? Why?
3. What did you observe when you mixed the barium hydroxide solution with the sulfuric acid solution? Use your observation to explain what could be happening.
4. Explain the difference between adding pure acetic acid to water and ethanol to water.
5. What do you infer from the conductivity of tap water?

EXTENSION

1. If the basement of your house is flooded with water, what should you be careful of?
2. What taste do perspiration and blood have in common? What might be in solution in perspiration and blood? Why should you be careful when working with electricity?

Agriculture and the Solubility of Salt

The most common soluble compound is sodium chloride (common salt). Sodium ions and chloride ions are the most abundant ions in sea water. How did these ions get into sea water? Over billions of years, rain has fallen on the rocks of the earth, dissolving the soluble ions, such as sodium and chloride, leaving behind the insoluble compounds. We tend to forget that this is an ongoing process, and it has now become a problem for agriculture in some regions.

The levels of the ions in river water are quite low and harmless, until the water is used for irrigation in hot, arid areas such as in the southwestern United States. There, water diverted from rivers is sprayed on the fields where it evaporates, leaving behind very small quantities of sodium chloride. But over time, as more and more water evaporates off the fields, large deposits of salt accumulate in the soil. Finally, when the salt concentration becomes too high, crops can no longer grow.

The simple solution to this problem is to apply enough water to soak the soil and carry away the salt, but in the American Southwest, water is scarce. Another solution is to develop salt-resistant crops through genetic engineering. The most economical, but least politically acceptable, solution is to stop growing water-demanding crops in the region and allow it to revert to its original semi-desert vegetation.

RELATED EXPERIMENT

Experiment 20 Solubilities of Salts

Experiment 20

Solubilities of Salts

INTRODUCTION

Some salts are soluble in water while others are not. In this experiment, you will mix combinations of anions and cations to try to identify what types of salts are insoluble and what types are soluble.

PROBLEM

Are there any patterns in the solubilities of salts?

APPARATUS AND MATERIALS

Per pair of students:

1 96-well plate
dropper bottle containing 0.50 mol/L solution of:
- chloride ions (sodium chloride)

dropper bottles containing 0.250 mol/L solutions of:
- sulfate ions (sodium sulfate)
- silver ions (silver nitrate)
- carbonate ions (sodium carbonate)
- barium ions (barium nitrate)
- chromate ions (potassium chromate)
- potassium ions (potassium nitrate)
- copper(II) ions (copper(II) nitrate)
- calcium ions (calcium nitrate)
- cobalt(II) ions (cobalt(II) nitrate)

SAFETY

Barium ions are toxic. Chromate ions are suspected carcinogens. However, in the concentrations used here, they present little risk. As an extra precaution, wash your hands thoroughly after the experiment. Silver ions stain skin and clothing.

PROCEDURE

1. Place 1 drop of each solution in the wells as indicated below.

Ion solution	Wells
chloride	A1 to F1
sulfate	A2 to F2
carbonate	A3 to F3
chromate	A4 to F4

2. Add 1 drop of each solution in the wells as indicated below.

Ion solution	Wells
silver	A1 to A4
barium	B1 to B4
potassium	C1 to C4
copper(II)	D1 to D4
calcium	E1 to E4
cobalt(II)	F1 to F4

3. Draw a table of anion against cation. List the names and formulas of the six positive ions in the first column. Write the names and formulas of the four negative ions in the first row. Record your observations as to which combinations give precipitates, colour changes, and so on.

ANALYSIS

1. Which metal cations form
   (a) insoluble chlorides?
   (b) insoluble sulfates?
   (c) insoluble carbonates?
   (d) insoluble chromates?

2. Chemists are interested in patterns in solubility.
   (a) Which anions *generally* form soluble salts?
   (b) Which anions *generally* form insoluble salts?

EXTENSION

1. Your teacher will provide you with some other sodium salts. Test these ion solutions using the same anions as in the Procedure.
2. Your teacher will provide you with other nitrate salts. Dissolve these salts. Test these ions using the same cations as in the Procedure.
3. Do these extra results reinforce your previous conclusions? Do they lead to any additional conclusions?

ENVIRONMENTAL APPLICATION

The best solution to the problem of waste chemicals is to minimize their production and to recycle them as much as possible. Suppose a factory produced some lead(II) ions in the waste water. Is it better to leave them in the water as it flows into a river or lake? Or is it better to precipitate the ion, filter off the insoluble solid, and bury it in a secure land fill site? If the latter, what inexpensive anion could you use to precipitate the lead(II)? Or is there a better solution to the problem?

Lakes and Solubility

All substances have a maximum solubility at a particular temperature. To put it another way, if you take a solution of any substance and allow the water to evaporate, the solution will become more and more concentrated and crystals of the substance will start to form. This process has happened on a large scale. Vast underground deposits of sylvinite (a mixture of potassium chloride and sodium chloride, otherwise known as potash) were formed when a gigantic prehistoric lake evaporated in what is now Saskatchewan. Over millions of years, layers of sand and clay covered the deposits. Today potash mining has great economic importance to Saskatchewan.

The same process happens today. It has become a serious problem in some places where rivers that feed lakes have been diverted for household, agricultural, or industrial use. Many people consider Mono Lake in California to be one of the most beautiful lakes in the world. Over the past few decades the rivers that feed it have been diverted to irrigate surrounding farmland. Because little fresh water is entering the lake, the salts in the lake are becoming more and more concentrated. If this continues, the lake may become a dry bowl of mineral deposits. Russia's Lake Baikal, the deepest lake in the world, is drying up because of diversions of the rivers that feed it. Because the lake is so big, its decreasing water level affects the climate. Also, as the salts in the water become more concentrated, the aquatic life in the lake will die. If the trend continues, Lake Baikal could become as salty—and as dead—as the Dead Sea. So, freshwater lakes are only fresh as long as sufficient amounts of water feed into them.

RELATED EXPERIMENT

Experiment 21 Solubility Curves

Experiment 21

Solubility Curves

INTRODUCTION

Different solutes dissolve differently in water. Also, temperature can have an effect on solubility. By knowing how temperature affects the solubility of a solute, a chemist can separate a solute from a solvent or even purify the solute. In this experiment, you will determine the solubility of a solute at different temperatures.

PROBLEM

How does the temperature affect the solubility of potassium nitrate?

APPARATUS AND MATERIALS

Per pair of students:

2 250-mL beakers
1 thermometer
1 jumbo pipette
1 balance
1 pipette funnel
1 1-mL graduated pipette
weighing paper
distilled water
0.4 g to 1.2 g potassium nitrate

PROCEDURE

1. Half-fill a 250-mL beaker with hot water (about 80°C), and half-fill another 250-mL beaker with cold water (below 20°C).
2. Measure and record the mass of the potassium nitrate sample that you have been given on a piece of weighing paper and then use the funnel to pour it carefully into the jumbo pipette as shown in the diagram on the next page left.

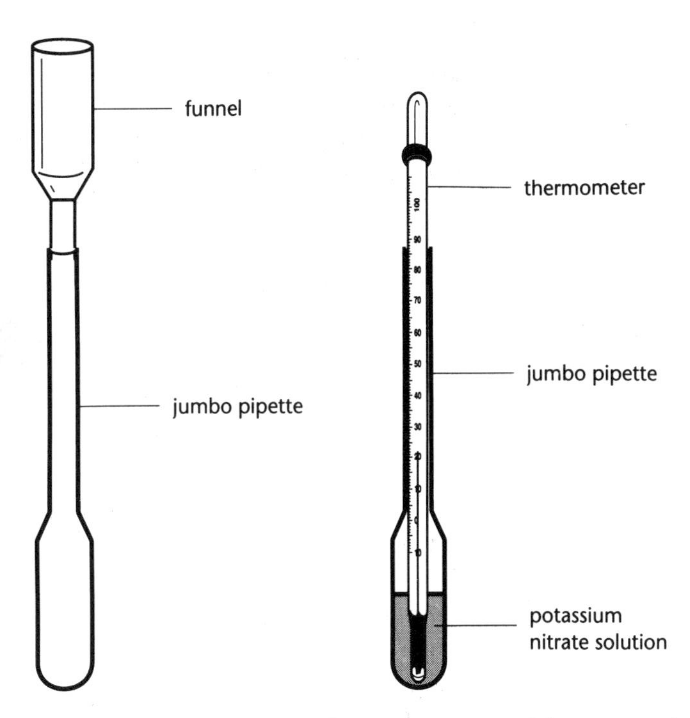

3. Remove the funnel. Using a graduated pipette, add 1.0 mL of distilled water to the jumbo pipette. Place a thermometer into the jumbo pipette as shown in the diagram above right.
4. Place the whole apparatus in the hot-water bath. Shake the solution until *all* the solute has dissolved. Remove the apparatus from the hot-water bath.
5. Place the apparatus in the cold-water bath until the *first* sign of crystals appears. *Do not let too many crystals form.* Note and record the temperature at which this happens.
6. Repeat Steps 4 and 5 once more to get a more accurate crystallization point.
7. Dispose of the solution into a beaker for recycling.

Optional: Use a computer spreadsheet to record and analyse your data.

ANALYSIS

1. Use the mass you dissolved in 1.0 mL of water to calculate the solubility of potassium nitrate in 100 mL of water.
2. Draw a graph of solubility (mass of potassium nitrate in grams per 100 g of water) against temperature (°C) on the *y*- and *x*-axes respectively. Use the values from other students to plot several points.
3. Draw the line of best fit through these points.

EXTENSION

1. (a) How could you recover most of the solute?
   (b) How could you recover the remaining solute?
2. What type of solution do the points on your line represent?
3. Use your graph to predict whether the following solutions are unsaturated, saturated, or supersaturated:
   (a) 38 g of potassium nitrate in 100 g of water at 23°C
   (b) 38 g of potassium nitrate in 100 g of water at 78°C.
4. From your graph, at what temperature would 115 g of potassium nitrate form a saturated solution in 100 g of water?

Experiment 22

Identifying Colourless Solutions

INTRODUCTION

In Experiment 18 you had several solutions that were colourless. In this experiment you will conduct further tests to try and categorize these solutions more specifically.

PROBLEM

How can you classify colourless solutions?

APPARATUS AND MATERIALS

Per pair of students:

1 96-well plate
1 24-well plate
1 stopper and delivery tube
5 gas-collecting pipettes
15 1-mL micro-tip pipettes
1 conductivity tester
24 toothpicks
1 burner
pH paper
red and blue litmus paper
6 1.0 mol/L solutions labelled A to F
grape juice
1 1-cm piece of magnesium ribbon
granular zinc
granular magnesium
sodium bicarbonate
calcium carbonate
limewater (calcium hydroxide)
distilled water

SAFETY

Limewater and solutions A to E are corrosive. Avoid contact with skin and clothing. Flush any contacted area with running water.

PRELAB ASSIGNMENT

Draw up a table with seven columns. In the first column, write these items:

name formula ions grape juice litmus paper
pH paper magnesium zinc sodium bicarbonate
calcium carbonate conductivity

At the top of the remaining six columns, write the headings A to F for your solutions. Record your observations in this table.

PROCEDURE

1. In the 96-well plate, use a 1-mL micro-tip pipette to place 4 drops of solution A in each of wells A1 to A8.
2. Repeat Step 1 for the other five solutions, inserting 4 drops into wells B1 to B8, C1 to C8, and so on, to F1 to F8.
3. Fill well A12 with distilled water.
4. Fill well G12 with limewater.
5. Add 2 drops of grape juice to each of wells A1 to F1. Record your observations.
6. Insert small pieces of red and blue litmus paper in each of wells A2 to F2. Record your observations.
7. Insert a small piece of pH paper in each of wells A3 to F3. Compare its colour to the pH disc and record the number.
8. Wet a toothpick in well A12 and use it to pick up some grains of magnesium. Place the magnesium in well A4. Repeat the procedure with the other solutions, inserting magnesium in wells B4 to F4. Record your observations.
9. Repeat Step 8 using zinc, and place it in wells A5 to F5, using a fresh toothpick.
10. Repeat Step 8 using sodium bicarbonate, and place it in wells A6 to F6, using a fresh toothpick.
11. Repeat Step 8 using calcium carbonate, and place it in wells A7 to F7, using a fresh toothpick.
12. To determine what gas was produced when magnesium was added to the different solutions, fill a gas-collecting pipette with tap water and place it into well A1 of the 24-well plate.

13. In the 24-well plate, fill well F1 about $\frac{1}{3}$ full of a fresh sample of a solution that produced a gas when magnesium was added. Then add 1 cm of magnesium ribbon. Immediately place the stopper and delivery tube on top of the well.

14. Wait about 15 s for any air to be displaced from the well. Put the gas-collecting pipette on top of the delivery tube as in the diagram below.

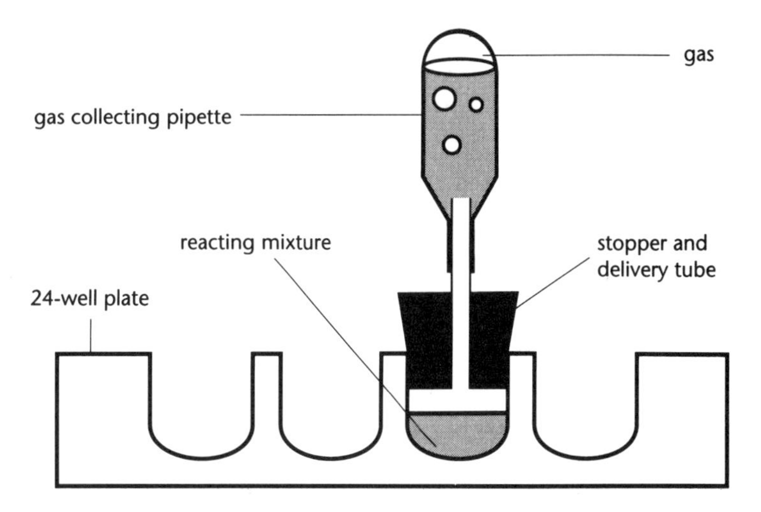

15. When the pipette is almost full of gas, lift it off and place it in well A1. The remaining water in the pipette will act as a plug for the collected gas.

16. Light a burner. Hold the pipette horizontally, about 2 cm from the flame, and quickly squirt the gas into the flame as shown in the diagram in Experiment 16. Record your observations.

17. Repeat Steps 12 to 16 using a different well for each of the other solutions that produced a gas when magnesium was added.

18. Repeat Steps 12 to 17 replacing magnesium with granular zinc. Use roughly the same volume of zinc as the magnesium ribbon.

19. Repeat Step 12 to 15 replacing magnesium with a small amount of sodium bicarbonate.

20. Use a 1-mL micro-tip pipette to draw up some limewater from well G12 to the shoulder of the pipette (see the diagram on the next page). *Keep the pressure on the bulb; do not release your fingers yet!* Place the tip of the pipette into the gas-collecting pipette. Now, slowly release the pressure on the pipette to allow the unknown gas to bubble through the limewater. Shake the gas/limewater mixture. Record your observations.

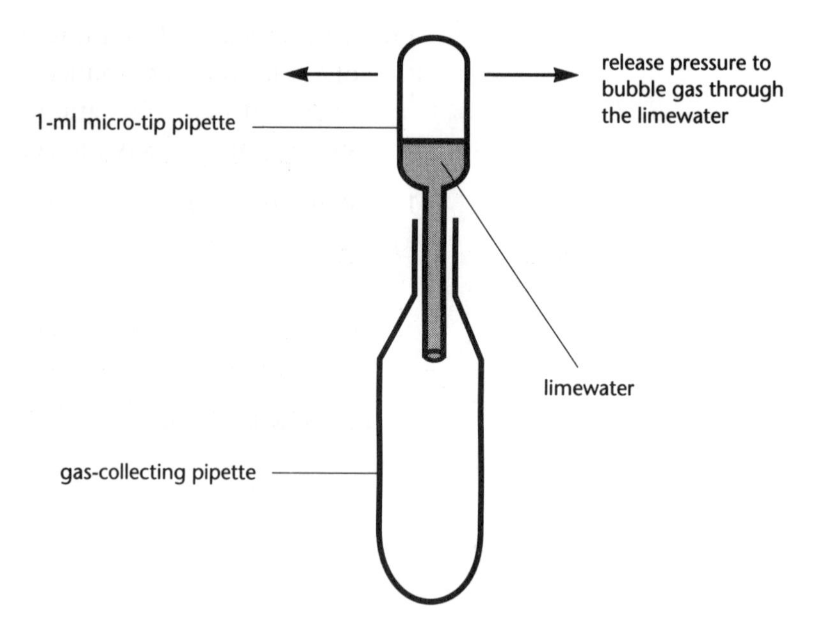

21. If there was no change, turn the apparatus upside down so that there is no limewater in the neck of the micro-tip pipette. Squeeze some of the gas back into the gas-collecting pipette and repeat Step 20.
22. Repeat Steps 19 to 21, using a different well for each solution that produced a gas when sodium bicarbonate was added.
23. Repeat Steps 19 to 22, replacing sodium bicarbonate with calcium carbonate, to test the gases produced by calcium carbonate.
24. Place the conductivity tester in well A8. Record your observations. Clean the tester tip and dry it.
25. Repeat Step 24 for each of wells B8 to F8.
26. Compare the six solutions and consider how to classify them.

ANALYSIS

1. How would you classify the solutions?
2. What do we call two of these classifications? What might solution F be?
3. Ask your teacher for the names and/or formulas of solutions A to F. What ion is common in one group of solutions? What ion is common in the other group of solutions?
4. What would happen if a solution from one group was mixed with a solution from the other group?

5. Other simple tests for these two groups are taste and feel. Why did you not conduct these tests?
6. In the kitchen at home, what are some common examples of solutions that belong to each of these two groups?

EXTENSION

Use pH paper to test several common solutions around the home. Also try a few solids dissolved in water. What can you say about most of the solutions you tested at home?

Where Did the Cadmium Come From?

One of the largest fields of employment for chemists is analytical chemistry. We are increasingly concerned about the traces of unusual elements and compounds in our foods. Analytical chemists devise methods for making precise measurements of trace materials. Once the chemists have tested the methods, they can use them to analyse samples.

One of the most interesting cases was the discovery by analytical chemists of high levels of cadmium in the Inuit of northern Quebec. Cadmium can cause a wide range of medical problems from high blood pressure to kidney failure. Initially, it was thought that the high concentrations of cadmium were due to the Inuit eating the organs of caribou. Analytical chemists had found the organs to be high in cadmium. However, after a ten-year ban on eating caribou organs, the Inuit's cadmium levels were as high as ever. Then it was noticed that the Inuit with high cadmium levels were all heavy cigarette smokers. A subsequent analysis of blood samples from cigarette smokers in Quebec City showed high levels of cadmium as well. Thus, the high cadmium levels had nothing to do with diet but were simply an example of one of the 3800 toxic compounds absorbed from cigarettes.

RELATED EXPERIMENTS

Experiment 23 Analysis of Vinegar
Experiment 24 Volumetric Acid-Base Titration

Experiment 23

Analysis of Vinegar

INTRODUCTION

Now that you have learned some properties of acids and bases you can apply them. One such application is the analysis of an acid or a base by the addition of a base or an acid to neutralize the acid or base (this technique is called titration). As most acids and bases are colourless, a colour-changing indicator has to be added to make the end point of the titration visible.

PROBLEM

How much acetic (ethanoic) acid is present in vinegar?

APPARATUS AND MATERIALS

Per pair of students:

1 24-well plate
3 1-mL micro-tip pipettes and labels
1 toothpick
1 balance
vinegar
0.500 mol/L solution of sodium hydroxide
phenolphthalein indicator

SAFETY

Sodium hydroxide is corrosive. However, in the concentration used here, it presents little risk. Wash any spills off your skin and clothing.

PROCEDURE

1. Fill a labelled pipette with vinegar. Measure and record its mass.
2. Fill another labelled pipette with a 0.500 mol/L solution of sodium hydroxide. Measure and record its mass.
3. Fill a well in the 24-well plate about $\frac{1}{3}$ full of vinegar from the pipette. Using a pipette, add 1 drop of phenolphthalein.

4. Add the sodium hydroxide solution to the vinegar, drop by drop while stirring continuously, until the indicator changes colour permanently. (The indicator will change colour from one drop to another.)
5. If you overshoot the end point, add some drops of vinegar to the well, and then continue carefully to add the sodium hydroxide.
6. Measure and record the mass of each pipette once the titration is complete. Calculate the masses of vinegar and sodium hydroxide used.
7. Repeat Steps 3 to 6 twice and record your data in a table.

Optional: Use a spreadsheet to record and analyse your data.

ANALYSIS

1. Use ratios to convert the masses of vinegar and sodium hydroxide to those which would have been used if 1 g of vinegar had been used.
2. Find the average mass of sodium hydroxide used.
3. Find the concentration of acetic acid in vinegar, in moles per litre.
4. Change this concentration to a mass percent. Compare this value to the one listed on the label of the vinegar bottle.
5. Why does the pink colour disappear quickly near the beginning of the titration and more slowly near the end?
6. If you added distilled water to the well with the vinegar, would it affect your results?
7. What should the pipettes be washed with prior to their use?
8. Why is it necessary to repeat the experiment at least once?

EXTENSION

1. Repeat the experiment with other vinegars such as pickling vinegar and cider vinegar.
2. Analyse various antacids for their ability to neutralize stomach acid.
3. Check the acidities of various carbonated soft drinks and find out what acids they contain.

Experiment 24

Volumetric Acid-Base Titration

INTRODUCTION

Chemists are not only concerned with what is present in a solution, but also how much. In this experiment, you will determine how much base is needed to neutralize a certain volume of two different acids, both having the same concentration.

PROBLEM

Do different acids of the same concentration require the same volume of base for neutralization?

APPARATUS AND MATERIALS

Per pair of students:

1 24-well plate
3 1-mL micro-tip pipettes
1 toothpick
0.10 mol/L solution of sodium hydroxide
0.10 mol/L solution of hydrochloric acid
0.10 mol/L solution of sulfuric acid
dropper bottle of phenolphthalein indicator solution

SAFETY

Sodium hydroxide, hydrochloric acid, and sulfuric acid are corrosive. However, in the concentrations used here, they present little risk. Wash any spills off your skin and clothing.

PROCEDURE

1. Your teacher will assign you a number between 15 and 25. This is the number of drops of each acid that you will use in this experiment. Always hold the pipette vertically when adding drops to ensure a consistent drop size.

2. In the 24-well plate, place 1 drop of phenolphthalein solution in each of wells A1 to A4 and D1 to D4.

3. Half-fill a pipette with hydrochloric acid solution. Add your assigned number of drops of hydrochloric acid solution to wells A1 to A4.

4. Half-fill a pipette with sulfuric acid solution. Add your assigned number of drops of sulfuric acid solution to wells D1 to D4.

5. Half-fill a pipette with sodium hydroxide solution. Add the sodium hydroxide solution, 1 drop at a time, to well A1, stirring while you add it. Continue adding the sodium hydroxide solution drop by drop until the indicator changes from colourless to where a pale pink colour persists. Record the number of drops of sodium hydroxide used.

6. Repeat Step 5 with wells A2 to A4. Calculate the mean value of the numbers of drops from the four measurements. (Ignore any value that is very different from the other values you got.)

7. Repeat Step 5 with wells D1 to D4. Calculate the mean value of the number of drops from the four measurements. (Ignore any value that is very different from the other values you got.)

ANALYSIS

1. Collect data from other students.

2. Plot the volume (drops) of hydrochloric acid solution added (y-axis) against the volume (drops) of sodium hydroxide (x-axis).

3. Plot the volume (drops) of sulfuric acid solution added (y-axis) against the volume (drops) of sodium hydroxide solution (x-axis).

4. (a) What do the two graphs indicate about the moles of hydrogen ion present in each acid?
   (b) Write a balanced chemical equation for each reaction to explain your conclusions in part (a).

Optional: Enter the class data on a spreadsheet and use a graphing program to print the graphs.

EXTENSION

1. Measure the mass of 50 drops of water and calculate the mass per drop. The densities of the solutions used in this experiment are close to 1.00 g/mL. Calculate the volume per drop. Determine the volume (in litres) of each reagent used. The concentration of the sodium hydroxide solution is 0.10 mol/L, and knowing the volume used, calculate the moles of base consumed. Using the balanced equation, calculate the moles of acid consumed. From the moles and volume of acid, find the concentration of each acid. Compare these values with the actual concentrations.

2. Repeat this experiment with other acids. Make sure their concentrations are 0.10 mol/L.

ENVIRONMENTAL APPLICATION

It is through the work of analytical chemists that we become aware of environmental problems. Whether or not analytical chemists use titrations or complex instruments, it is their job to analyse samples of the air that we breathe and the water that we drink. Only by chemical analysis can we be sure that the air and water meet the specifications of purity defined by law.

How "Scientific" Is Science?

Although science is portrayed as being a logical discipline, this is not always true. Scientists certainly use logic when they devise and perform experiments. But when they interpret their findings, they may use both logic and intuition. When a particular scientific viewpoint has been accepted, many scientists stubbornly cling to that idea, even though a new approach may make more sense. In 1803, when the atomic theory was first proposed, many famous chemists passionately denounced the theory. Years later, and only when the evidence became overwhelming, was the atomic theory generally accepted.

Sometimes the evidence for a scientific idea is inconclusive. One of the most interesting examples is the claim put forward by Dr. Linus Pauling that large doses of vitamin C can inhibit the common cold virus as well as other illnesses. Pauling had won two Nobel prizes: one for inorganic chemistry and one for peace. Thus, his brilliance was well established. However, his research on vitamin C was in biochemistry, a field different from his expertise. Other scientists have tried to repeat his work with varying success. In every experiment that produced contrary results, Pauling claimed to find a flaw in the procedure. On the other hand, some experiments do appear to support his ideas. Which is correct? When human beings are being experimented on, definite, clear-cut results are difficult to obtain. At the moment, all one can say is that more research is needed—and it must be conducted by people who are neither "pro" nor "anti" Pauling's ideas.

RELATED EXPERIMENT

Experiment 25 Vitamin C Analysis

Experiment 25

Vitamin C Analysis

INTRODUCTION

You should now be proficient at performing titrations. In this experiment, you will use this skill to check the vitamin C level in apple juice and, if time permits, in other juices and drinks. The starch in the solution acts as an indicator. During the titration, the vitamin C will react with the iodine solution to keep the solution colourless. When all the vitamin C has reacted with the iodine, the excess iodine will react with the starch in the solution to form a bluish-black colour. Thus the end point is indicated by starch.

PROBLEM

How does the vitamin C content vary in different samples of apple juice?

APPARATUS AND MATERIALS

Per pair of students:

1 24-well plate
3 1-mL micro-tip pipettes and labels
1 toothpick
starch solution
iodine solution
various apple juices including
    fresh
    from concentrate
    apple drink (as opposed to apple juice)
standard vitamin C solution (0.100%)

SAFETY

Iodine is corrosive and can stain skin and clothing. Remove iodine stains by using some vitamin C solution!

PRELAB ASSIGNMENT

Bring from home a variety of apple juices and apple drinks including fresh apple juice. Leave a sample of each out of the refrigerator to age for a few days.

PROCEDURE

PART I—STANDARDIZATION OF THE IODINE SOLUTION

1. Fill a labelled pipette with standard vitamin C solution. Measure and record its mass.
2. In the 24-well plate, fill a well $\frac{1}{3}$ full with the standard vitamin C solution.
3. Measure and record the mass of the pipette again. Calculate the mass of standard vitamin C solution in the well.
4. Use a pipette to add 1 drop of starch indicator to the well.
5. Fill another labelled pipette with iodine solution. Measure and record its mass.
6. Add the iodine solution drop by drop to the vitamin C solution, stirring continuously with a toothpick, until the solution turns permanently bluish black.
7. Measure and record the mass of the pipette containing the iodine solution. Calculate the mass of iodine solution used.
8. Repeat Steps 2 to 7.

Optional: Use a computer spreadsheet to record and analyse your data.

PART II—JUICE ANALYSIS

9. Repeat Steps 1 to 8 for each juice you provided—both the refrigerated and aged samples. In each case, replace the vitamin C solution with a fruit juice.

ANALYSIS

PART I

1. For each trial, calculate the mass of vitamin C solution that would react with 1.00 g of iodine solution. Find the average value of the mass.
2. The vitamin C solution is 0.100%. What mass (in milligrams) of vitamin C is in 1.00 g of vitamin C solution?
3. How much vitamin C reacts with 1.00 g of iodine solution?

PART II

4. If a normal serving of apple juice is 100 mL (100 g), calculate how much iodine solution would be needed to neutralize each sample in the experiment if it contained 100 mL.

5. Use your conversion factor from Step 3 in the Analysis to find the mass (in milligrams) of vitamin C in each sample of juice.
6. (a) How does aging affect the vitamin C content of juices?
   (b) How does the vitamin C content of fruit juice compare with fruit drink?
   (c) How does the vitamin C content of fresh fruit juice compare with the fruit juice made from concentrate?

EXTENSION

1. What is the recommended daily intake of vitamin C?
2. Why is vitamin C important in our diet?
3. Compare the vitamin C content of other types of fruit juices and/or drinks.
4. Research the work of Linus Pauling.
5. Try to analyse the vitamin C content of vegetables. What problems might you encounter when attempting to conduct such an analysis?

Experiment 26

Actual and Theoretical Yields

INTRODUCTION

This experiment combines many concepts that you have studied in chemistry: the gas laws, balanced equations, stoichiometry, acids, limiting reagents, and percentage yield.

In industry it is important to know how efficient a reaction is. By comparing theoretical yield to actual yield, chemists can calculate the percentage yield. From this value, they can tell if something is wrong in the reaction process.

In this experiment, you will react magnesium with hydrochloric acid to produce a gas. After calculating theoretical and actual yields of this gas, you can find the percentage yield.

PROBLEM

What is the percentage yield of gas produced in the reaction of magnesium with hydrochloric acid?

APPARATUS AND MATERIALS

Per pair of students:

1 50-mL beaker
1 10-mL syringe (wide opening)
1 pair of tweezers
1 barometer
1 15-cm ruler
1 cleaned magnesium ribbon less than 1 cm long
1 dropper bottle of 1.00 mol/L hydrochloric acid solution

SAFETY

Hydrochloric acid is corrosive. Avoid contact with skin and clothing. Flush any contacted area with running water.

PRELAB ASSIGNMENT

Write a balanced equation for the reaction of magnesium with hydrochloric acid.

PROCEDURE

1. Measure accurately the length of a piece of magnesium ribbon.
2. Fill the syringe with 10.0 mL of the hydrochloric acid solution. Record the volume.
3. Hold the syringe over the beaker as shown in the diagram below, and use tweezers to place the magnesium in the open end of the syringe. Immediately release the magnesium. The solution will be pushed out by the gas into the beaker.

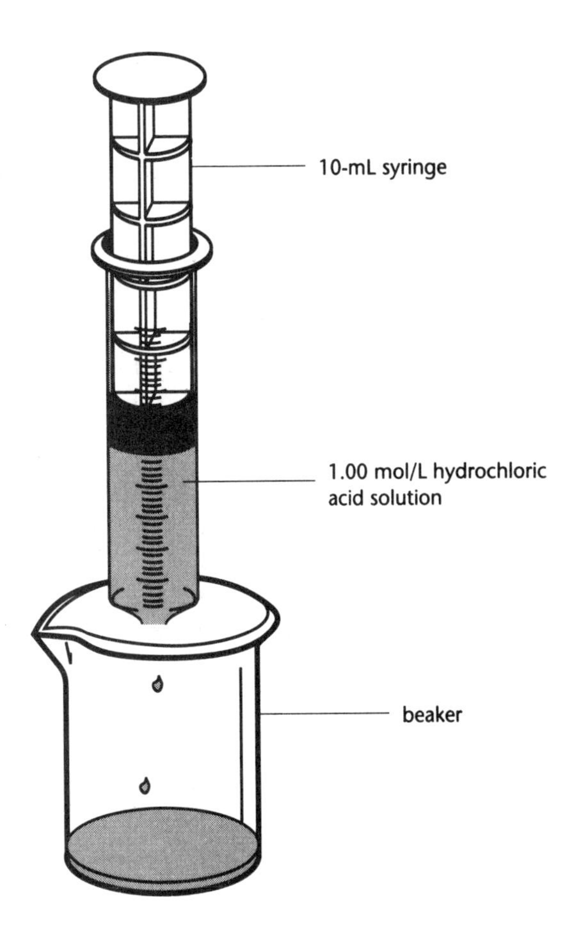

4. Hold the syringe upright until all the magnesium has reacted. Wait for a few minutes until the contents of the syringe have cooled to room temperature. Record the final volume of the liquid in the syringe.
5. Record the barometric pressure and room temperature.
6. Return the acid solution to your teacher.

ANALYSIS

1. Ask your teacher for the mass and length of a longer piece of magnesium ribbon. Use this information to find the mass of your piece of magnesium.
2. Use the equation from the Prelab Assignment to predict the volume, at standard temperature and pressure, of the gas produced from the mass of magnesium you used. This is the theoretical yield.
3. You collected your gas over water (the acid solution is mostly water) so the gas is "wet." Consult the appropriate table to correct your pressure reading and calculate the pressure of "dry" gas.
4. Convert the volume of the gas you collected in the laboratory to that which would have collected at standard temperature and pressure. This is the actual yield.
5. Calculate the percentage yield.

$$\text{Percentage yield} = \frac{\text{Actual yield}}{\text{Theoretical yield}} \times 100\%$$

EXERCISES

1. (a) What gas was produced?
   (b) How could you have tested for its identity?
2. (a) How close to 100% is your value for the percentage yield?
   (b) How accurate was your value for the mass of magnesium?
   (c) What effect could this mass measurement have on your answer?
   (d) What else might affect the percentage yield?

Fossil Fuels and Exothermic Chemical Reactions

Many, but not all, chemical reactions are accompanied by the release of heat energy. Some of these exothermic chemical reactions are essential for our survival. They produce the energy we use for cooking, warmth, and transportation. In the developed countries, we burn coal, oil, or natural gas to produce these exothermic reactions from which we get our energy. In some of these reactions, the energy is produced directly. For example, we use gasoline to power the internal combustion engines in our cars. In other reactions, the energy we use is produced indirectly. For example, we burn oil and coal in generating plants to provide us with electrical energy. But these sources of energy are all fossil fuels, laid down hundreds of millions of years ago, and the reserves will become exhausted in the next century. Thus, the "fossil fuel age" will be the shortest epoch in the history of humanity. Fossil fuels are not just sources of energy. They are complex molecules that can be used much more effectively as raw materials for the production of pharmaceuticals and plastics. So it is imperative that we reduce our dependence on fossil fuels as sources of exothermic reactions and instead use non-destructive sources of energy, such as the wind-power electrical generators that have proved so successful in Denmark and California.

RELATED EXPERIMENTS

Experiment 27 Heats of Reaction
Experiment 28 Heat of Combustion of Magnesium

Experiment 27

Heats of Reaction

INTRODUCTION

One of the simplest ways to measure the amount of heat produced by an exothermic reaction is to perform a reaction in solution in a calorimeter (a "heat-collecting" container).

In this experiment, you will perform three related reactions and then compare the heats of reaction and the appropriate reaction equations.

PROBLEM

How do three heat-of-reaction values relate to each other?

APPARATUS AND MATERIALS

Per pair of students:

1 10-mL graduated cylinder
1 balance
1 thermometer
3 25-mL plastic cups
1 small elastic band
0.8 g solid sodium hydroxide
10.0 mL 1.00 mol/L solution of sodium hydroxide
20.0 mL 0.500 mol/L solution of hydrochloric acid
10.0 mL 1.00 mol/L solution of hydrochloric acid

SAFETY

Solid sodium hydroxide is very corrosive. Handle with care. The sodium hydroxide and hydrochloric acid solutions are also corrosive. Avoid contact with skin and clothing. Flush any contacted area with running water.

PRELAB ASSIGNMENT

Write a net ionic equation for each reaction:

(a) Solid sodium hydroxide dissolves in water to form ions.
(b) Solid sodium hydroxide reacts with hydrochloric acid solution to form water and aqueous sodium chloride.
(c) Aqueous sodium hydroxide reacts with hydrochloric acid solution to form water and aqueous sodium chloride.

PROCEDURE

PART I—REACTION OF SOLID SODIUM HYDROXIDE AND WATER

1. Place one plastic cup inside the other using the elastic band as a spacer, as shown in the diagram below. This forms a calorimeter.

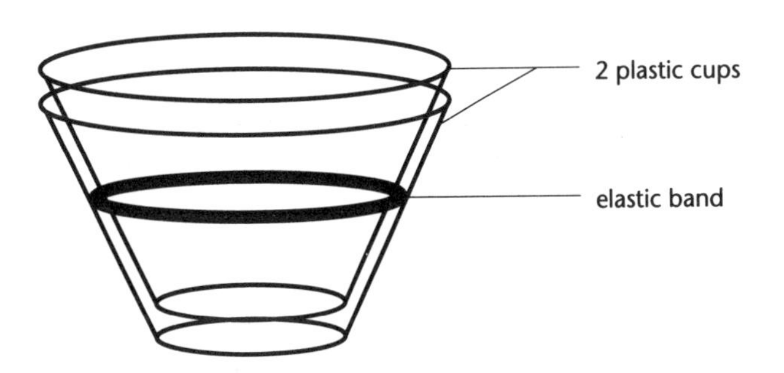

2. Using the 10-mL graduated cylinder, pour exactly 20.0 mL of tap water into the calorimeter. Stir with the thermometer and record the temperature.

3. Measure about 0.4 g of solid sodium hydroxide. Record the exact amount you measured to the nearest thousandths of a gram.
4. While stirring the water with the thermometer, add the solid sodium hydroxide. Record the maximum temperature reached by the solution.

5. Dispose of the solution as instructed by your teacher and clean the calorimeter.

PART II—REACTION OF SOLID SODIUM HYDROXIDE AND HYDROCHLORIC ACID SOLUTION

6. Repeat Steps 2 to 5 but replace the tap water with 20.0 mL of a 0.500 mol/L solution of hydrochloric acid.

PART III—REACTION OF SODIUM HYDROXIDE SOLUTION AND HYDROCHLORIC ACID SOLUTION

7. Pour exactly 10.0 mL of a 1.00 mol/L solution of hydrochloric acid into the calorimeter.
8. In another single plastic cup, measure exactly 10.0 mL of a 1.00 mol/L solution of sodium hydroxide.
9. Record the temperature of each solution. Clean your thermometer before each measurement.
10. Mix the two solutions in the calorimeter. Record the maximum temperature reached.

Optional: Use a computer spreadsheet to record and analyse your data.

ANALYSIS

1. Calculate the temperature change of each solution.
2. Since these solutions are dilute, we assume that their densities are 1.0 g/mL. Calculate the heat absorbed by each solution.
3. For each reaction, calculate the number of moles of sodium hydroxide used.
4. For each reaction, calculate the amount of heat evolved per mole of sodium hydroxide.
5. Compare your results with those of other students. Find a simple mathematical relationship among the three values you found in Step 4.
6. Now compare the three net ionic equations for the three reactions in the same way as you did in Step 5.

EXERCISES

1. (a) Write the three equations, including the heat term in each equation.
   (b) Rewrite each equation using ΔH notation.
2. Suppose you had used ten times as much sodium hydroxide in the first reaction.
   (a) What effect would this have on the temperature change and, hence, the amount of heat produced?
   (b) What effect would this have on the heat evolved per mole of sodium hydroxide?
3. What would happen to the temperature if an endothermic reaction had occurred?

EXTENSION

How could the energy be found in foods such as cereals or peanuts?

Experiment 28

Heat of Combustion of Magnesium

INTRODUCTION

You have found the heat of a reaction experimentally and learned how to find the heat of a reaction when that heat cannot be measured in the laboratory. You may have seen burning magnesium in an earlier science course and noticed the tremendous amounts of heat and light energy produced.

In this experiment, you will produce two related reactions of magnesium and its compounds to find indirectly the heat of combustion of magnesium.

PROBLEM

What is the heat of combustion of magnesium?

APPARATUS AND MATERIALS

Per pair of students:

1 thermometer
2 25-mL plastic cups and elastic band (calorimeter)
1 25-mL graduated cylinder
1 balance
1 piece of emery paper
0.1 g magnesium ribbon
0.2 g magnesium oxide
40.0 mL of 1.00 mol/L solution of hydrochloric acid

SAFETY

Hydrochloric acid is corrosive. Avoid contact with skin and clothing. Flush any contacted area with running water.

PROCEDURE

PART I—REACTION OF MAGNESIUM OXIDE AND HYDROCHLORIC ACID SOLUTION

1. Measure about 20.0 mL of a 1.00 mol/L solution of hydrochloric acid into the calorimeter. Record the amount you used.
2. Measure and record the temperature of this solution.
3. Measure about 0.2 g of magnesium oxide and add it to the calorimeter. Record the amount you used.
4. Stir the contents of the calorimeter carefully and record the maximum temperature reached.
5. Dispose of the solution as directed by your teacher. Clean the apparatus.

PART II—REACTION OF MAGNESIUM AND HYDROCHLORIC ACID SOLUTION

6. Repeat Steps 1 to 5 but replace the magnesium oxide with 0.1 g of magnesium ribbon. Clean the magnesium ribbon with emery paper before use.

Optional: Use a computer spreadsheet to record and analyse your data.

ANALYSIS

1. Write the reaction equation for the reaction in Part I and in Part II.
2. Calculate the temperature change for each reaction.
3. Since you are using dilute solutions, assume that their densities are 1.0 g/mL. Find the quantity of heat absorbed by each solution.
4. (a) Calculate the number of moles of magnesium oxide used.

   (b) Calculate the number of moles of magnesium used.
5. Find the heat per mole for each reaction.
6. Use the two reaction equations and their heat values (ΔH) along with an equation from a ΔH table to find the heat of combustion of magnesium.

EXERCISES

1. (a) Compare your value for the heat of combustion of magnesium to the accepted value.

   (b) Give two reasons why your value may be different.

Experiment 29

Effect of Surface Area on the Rate of Reaction

INTRODUCTION

In this and the next two experiments, you will study the factors that affect the rate of a reaction. All of these factors have many applications in your daily lives: in cooking, in keeping food fresh, and in fighting fires, to name a few.

When a solid is broken into smaller pieces, its surface area increases. In this experiment, you will observe how different-sized pieces of calcium carbonate react with hydrochloric acid.

PROBLEM

What effect will increased surface area have on the rate of a reaction?

APPARATUS AND MATERIALS

Per pair of students:

1 24-well plate
1 1-mL micro-tip pipette
1 balance
1 pestle and mortar
1 stopwatch or clock with a second hand
2 similar-sized marble chips (calcium carbonate)
1 dropper bottle of 4.0 mol/L solution of hydrochloric acid

SAFETY

Hydrochloric acid is very corrosive. Handle with care. Avoid contact with skin and clothing. Flush any contacted area with running water.

PRELAB ASSIGNMENT

Make a table with these three headings: Time (s), Mass (g) whole piece calcium carbonate, Mass (g) crushed piece calcium carbonate.

Draw enough lines in the table for the Time column to go from 0 s to 120 s, in 10 s intervals.

PROCEDURE

1. Fill the pipette with a 4.0 mol/L solution of hydrochloric acid.
2. Place a 24-well plate on the balance and place the pipette in a well.

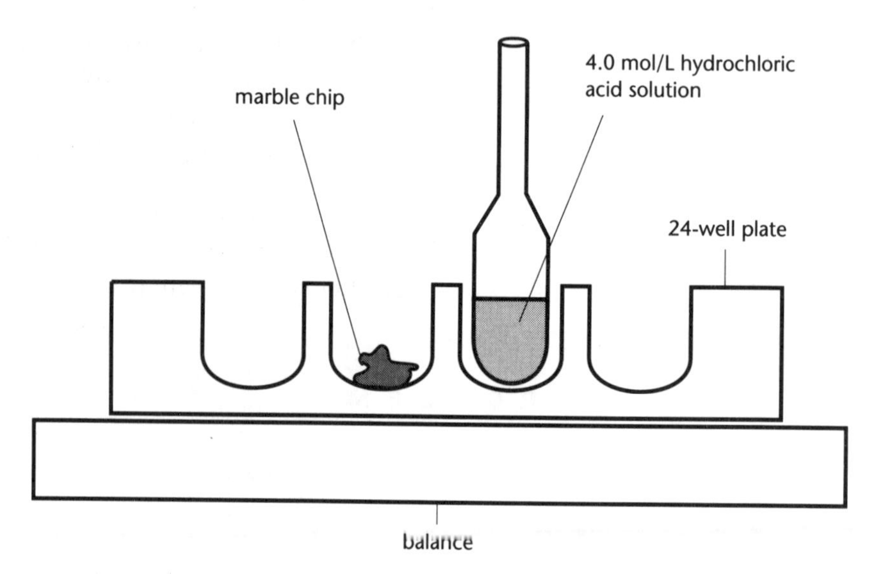

3. Put one marble chip in an adjacent well and zero the balance.

4. When you are ready to start timing, quickly squirt the hydrochloric acid solution onto the marble chip until the well is $\frac{1}{3}$ full. Place the pipette in its well, and record the mass every 10 s for 2 min.
5. Clean the apparatus.
6. Break a second marble chip into about 20 smaller pieces with a pestle and mortar. *Do not grind the chip into a powder.*
7. Repeat Steps 1 to 5, replacing the marble chip with the crushed chip.

ANALYSIS

1. On the same grid, plot mass loss (y-axis) against time (x-axis) for the two reactions. Draw a smooth curve through each set of points.
2. For each graph, measure the average rate of each reaction (the slope of the graph) in the first 20 s and the last 20 s.
3. Suppose you continued to measure the mass loss beyond 2 min. and plotted the values. What would the lines on your graph look like?
4. Write an equation for the reaction.

EXTENSION

1. (a) When you are lighting a campfire, what size of wood is best to start it?
   (b) What size of wood would you use to keep the fire going overnight?
2. In pioneer days, how was cheese kept to slow down its spoilage?
3. Why do we chew food before swallowing it?
4. Use collision theory to explain the effect of surface area on reaction rate.
5. Dust explosions are very dangerous in coal mines and grain elevators. Why?
6. Research the Westray mining disaster in Nova Scotia in 1992, and its possible causes.

Experiment 30

Effects of Concentration and Temperature on the Rate of Reaction

INTRODUCTION

Two more factors affecting the rate of a reaction are concentration and temperature. Each of these effects can be observed in the reaction between solutions of potassium iodate and sodium metabisulfite. Although this is a rather complex reaction involving the following steps:

1) $IO_3^-(aq) + 3HSO_3^-(aq) \longrightarrow I^-(aq) + 3SO_4^{2-}(aq) + 3H^+(aq)$

2) $5I^-(aq) + 6H^+(aq) + IO_3^-(aq) \longrightarrow 3I_2(aq) + 3H_2O(l)$

3) $I_2(aq) + HSO_3^-(aq) + H_2O(l) \longrightarrow 2I^-(aq) + SO_4^{2-}(aq) + 3H^+(aq)$

4) $I_2(aq)$ + starch $\longrightarrow$ blue colour

you can observe the effects that both concentration and temperature have on the rate of the reaction by timing the colour change of the reaction.

PROBLEM

What effects will a change in concentration and a change in temperature have on the rate of a reaction?

APPARATUS AND MATERIALS

Per pair of students:

2 96-well plates
2 1-mL micro-tip pipettes
1 thin-stem pipette (cut off)
1 gas-collecting pipette
1 thermometer
1 stopwatch or clock with a second hand

warm- and cold-water baths
dropper bottles of:
0.020 0 mol/L solution of potassium iodate
0.001 00 mol/L solution of sodium metabisulfite/starch/acid
distilled water

SAFETY

Sodium metabisulfite is corrosive. Wash any spills off your skin and clothing. Flush any contacted area with running water. Caution: The sodium metabisulfite solution and the sulfur dioxide gas that can be produced can cause an allergic reaction in some people.

PROCEDURE

PART I—THE EFFECT OF CONCENTRATION ON REACTION RATE

1. In a 96-well plate, place drops of potassium iodate solution in rows A, C, and E and dilute with distilled water according to the following table and as shown in part (a) of the diagram below.

Column	1	3	5	7	9
iodate water	1 4	2 3	3 2	4 1	5 0

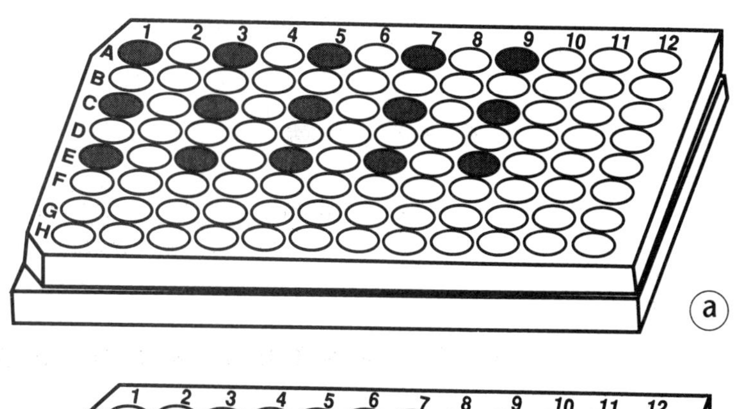

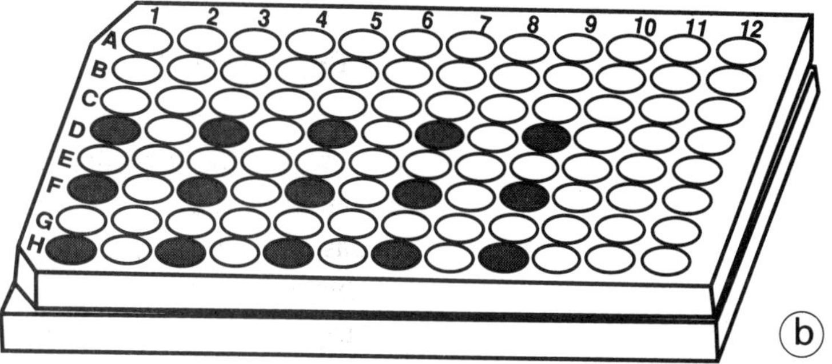

2. In the other 96-well plate, place 5 drops of sodium metabisulfite solution in rows D, F, and H, and columns 1, 3, 5, 7, and 9 as shown in part (b) of the diagram on the previous page.
3. Carefully invert the second plate on top of the first plate. No, the solutions won't fall out! When you are ready to time the reactions, quickly drop the 2-plate "sandwich" about 50 cm, and stop abruptly. This will shake down all the solutions from the top plate into the lower plate. This produces three sets of reactions for each of the five different concentrations of potassium iodate solutions. Record only one time for each concentration.
4. For each set of three reactions at each concentration, record the time the second solution turns blue (this is the median value).
5. Clean the apparatus.

PART II—THE EFFECT OF TEMPERATURE ON REACTION RATE

You will control the concentrations by using 5 drops of each solution in separate 1-mL micro-tip pipettes.

6. Place 5 drops of potassium iodate solution into the thin-stem pipette bulb using a 1-mL micro-tip pipette.
7. Place 5 drops of sodium metabisulfite solution into a gas-collecting pipette using a 1-mL micro-tip pipette.
8. Insert the thin-stem pipette into the gas-collecting pipette and, when you are ready to start timing, squirt the potassium iodate solution into the sodium metabisulfite solution.
9. Record how long it takes for the colour change to occur. Also record the room temperature.
10. Clean the apparatus.

    When you are comfortable with this technique, repeat the reaction about 10°C above and 10°C below room temperature using a warm- and a cold-water bath outlined in the Steps below. Ensure your solutions are never heated to 40°C.

11. Repeat Steps 6 to 10 but place the two solutions in a warm-water bath for about 5 minutes.
12. Mix as before but keep the mixture in the warm-water bath until the reaction has occurred. See the diagrams on the next page.

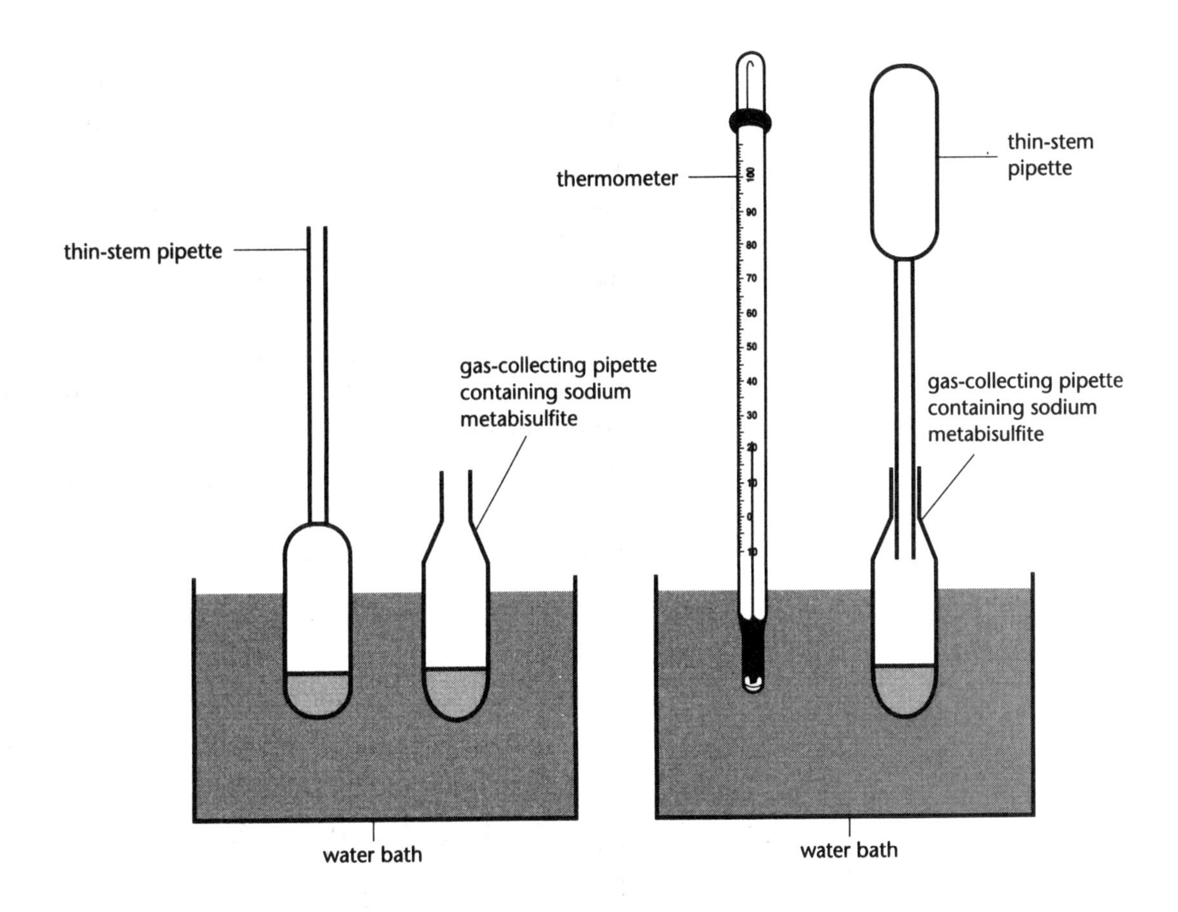

13. Record the temperature of the warm-water bath as shown in the above diagram.

14. Repeat the Steps 11 to 13 using a cold-water bath.

ANALYSIS

PART I

1. Calculate the concentration of potassium iodate in each well at the moment it was mixed with the sodium metabisulfite solution.

2. Draw a graph of concentration (y-axis) against time (x-axis).

PART II

3. Draw a graph of temperature (y-axis) against time (x-axis).

EXTENSION

1. Use the results of this experiment to suggest how you could prevent food (for example, cheese) from going bad.
2. Why are fire doors used in schools? Relate your answer to this experiment.
3. Describe two ways to extinguish a fire.
4. What simple effective way is there to extinguish a fire in a pot on the stove? Explain.

Biological Catalysts and Rate of Reaction

Every chemical reaction takes place at a certain rate at a certain temperature. Some reactions are very fast, such as the decomposition of the explosive TNT, while other reactions are very slow. At any given time, millions of different chemical reactions happen in our bodies. Some of them are quite slow. For example, the digestion of our food (the breakdown of complex molecules into simple sugars and amino acids) takes several hours. This is fortunate, for we wouldn't want to swallow our food and have the reaction take place in seconds. If it did, our stomachs would probably explode from the vigorous reaction.

Some chemical reactions in our bodies would happen extremely slowly if it weren't for the presence of catalysts. These catalysts can increase the rate of reaction by a factor of millions or billions. Biochemists call these catalysts enzymes. Alcohol dehydrogenase is one such enzyme. It speeds up the breakdown of alcohols into water and carbon dioxide. If we did not have this enzyme, one alcoholic drink would leave a person inebriated for days. Why do we possess this enzyme? One explanation is that our prehistoric ancestors could not reach much of the fruit on trees so they ate the fruit that had fallen to the ground. The fallen fruit had probably started to ferment, producing alcohol. Those individuals who could metabolize the alcohol faster would have an obvious advantage over those who could not.

RELATED EXPERIMENT

Experiment 31 Effect of a Catalyst on the Rate of Reaction

Experiment 31

Effect of a Catalyst on the Rate of Reaction

INTRODUCTION

Catalysts are very important chemicals that we rarely appreciate because we do not write them in equations and they are not often mentioned in chemical reactions. Our bodies could not function without a wide range of catalysts, and they are also used industrially to make gasoline, synthetic fibres, and most of the drugs we use.

In this experiment, you will identify a catalyst and observe how it works when hydrogen peroxide decomposes.

PROBLEM

Do marble and pyrolusite rock behave as catalysts?

APPARATUS AND MATERIALS

Per pair of students:

1 24-well plate
1 stopper and delivery tube
1 gas-collecting pipette
1 graduated jumbo pipette
1 pair of tweezers
1 balance
1 hair dryer
1 stopwatch or clock with second hand
1 marble chip
1 piece of pyrolusite rock
dropper bottles of:
  6% hydrogen peroxide solution
  acetone
distilled water

SAFETY

Acetone is flammable. Make sure there are no open flames in the laboratory. Hydrogen peroxide is very corrosive. Handle with care. Avoid contact with skin and clothing. Flush any contacted area with running water.

PROCEDURE

1. Use the graduated jumbo pipette to place 2 mL of hydrogen peroxide solution into a well in the 24-well plate.
2. Place the stopper and tube on this well, then place the gas-collecting pipette, filled with tap water, on the tube as shown in the diagram below.

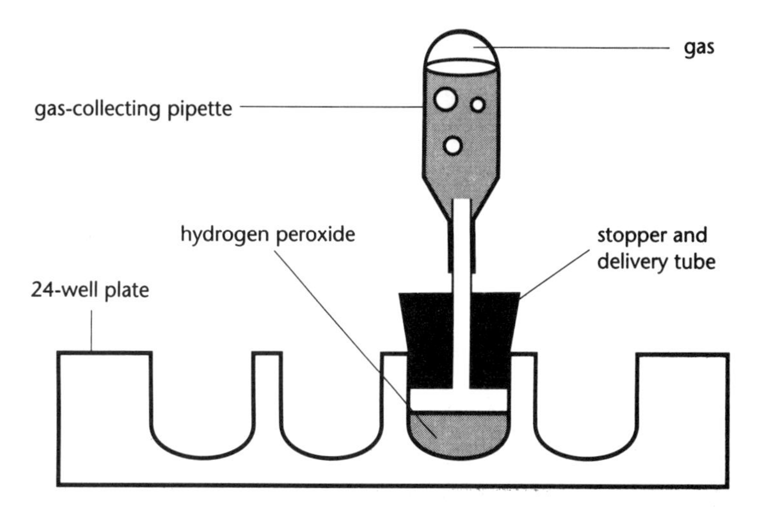

3. Record the number of gas bubbles produced every 10 s for 60 s.
4. Using tweezers, place a marble chip on the balance and record its mass.

NOTE: Never touch the marble or pyrolusite chips. Oils on your skin can be transferred to the chips. These oils can affect the results of the experiment. Always use tweezers to add the chips to the well.

5. Repeat Step 1, using fresh hydrogen peroxide in a new well.
6 When you are ready to start timing, add the marble chip to the well and carefully but quickly repeat Steps 2 and 3.
8. After 60 s, remove the marble chip, wash it with distilled water and then with acetone. Dry it with a hair dryer and then measure and record its mass.
9. Repeat Steps 4 to 8, replacing the marble chip with a piece of pyrolusite rock.

ANALYSIS

1. On the one graph, plot the number of bubbles (y-axis) against time (x-axis) for each of the three reactions.
2. (a) Did the marble chip and/or pyrolusite rock speed up the decomposition of the hydrogen peroxide solution?
   (b) Is marble a catalyst?
   (c) Is pyrolusite rock a catalyst?

EXERCISES

1. (a) What gas was produced in this decomposition reaction?
   (b) How could you test for the gas?
2. Write an equation for the reaction.

EXTENSION

1. What are the catalysts in our body called?
2. (a) Why is a catalytic converter used in cars?
   (b) What element is used in catalytic converters? How does the cost of this element compare with the cost of other elements?

Experiment 32

Rate Law for a Reaction

INTRODUCTION

Although a general understanding of the various factors affecting the rate of a reaction can be useful, it is more advantageous to know quantitatively what happens in a reaction. This can be determined from the rate law.

In this experiment, you will determine the quantitative effect that doubling the concentration has on the rate of decomposition of hydrogen peroxide.

$$H_2O_2(aq) \longrightarrow H_2O(l) + \frac{1}{2}O_2(g)$$

PROBLEM

What is the rate law for the decomposition of hydrogen peroxide?

APPARATUS AND MATERIALS

Per pair of students:

1 24-well plate
1 stopper and delivery tube
1 gas-collecting pipette
2 graduated jumbo pipettes
1 stopwatch or clock with a second hand
1 pair of tweezers
1 piece of pyrolusite rock
distilled water
dropper bottle of 6% hydrogen peroxide solution

SAFETY

Hydrogen peroxide is very corrosive. Handle with care. Avoid contact with skin and clothing. Flush any contacted area with running water.

PROCEDURE

1. Use a graduated jumbo pipette to place 2.0 mL of hydrogen peroxide solution into a well on the 24-well plate.
2. Fill the gas-collecting pipette with tap water.
3. Using tweezers, add the pyrolusite rock to the well with the hydrogen peroxide solution. Immediately insert the stopper and tube, then place the gas-collecting pipette on top of the tube as shown in the diagram below.

NOTE: Never touch the pyrolusite rock. Oils from your skin can be transferred to the rock. These oils can affect the results of the experiment. Always use tweezers to add the rock to the well.

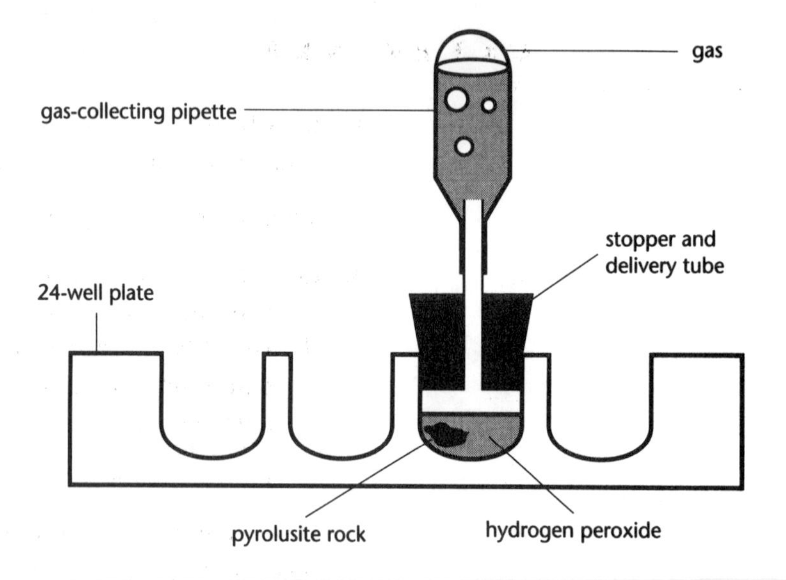

4. After 10 s, count and record the number of gas bubbles produced in approximately 20 s (for example, 12 bubbles/21 s).
5. Repeat Step 4 to get a second reading so you have recorded about 40 s worth of bubbles.
6. Dispose of the solutions as instructed by your teacher. Rinse the well plate.
7. Repeat Steps 1 to 6, using different wells and using each of these dilutions of the hydrogen peroxide solution.

hydrogen peroxide	1.5 mL	1.0 mL	0.5 mL
distilled water	0.5 mL	1.0 mL	1.5 mL

Optional: Use a computer spreadsheet to record and analyse your data.

ANALYSIS

1. Calculate the rate (bubbles per second) for each reading. Find the average rate for each of the four concentrations.
2. Calculate the concentrations of the hydrogen peroxide solutions.
3. Graph the average rate (y-axis) against the concentration of the hydrogen peroxide solution (x-axis).
4. Is the graph a straight line? If it is, there is a direct relationship between rate and concentration. This is the rate law. Write a formula that describes the straight line.

EXERCISES

1. If each reaction were allowed to continue, what would eventually happen to its rate? Why?
2. What is the purpose of the pyrolusite rock?
3. (a) Measure the slope of the line and rewrite the rate law with this value of the rate constant included.
   (b) Use this equation to find how many bubbles of oxygen would be produced (using the undiluted hydrogen peroxide) in

   i) 10 s ii) 60 s and iii) 300 s.
4. Suppose you need 120 bubbles of oxygen to conduct an experiment. Using the undiluted hydrogen peroxide, about how long would it take to produce this amount?

Where Is the Carbon Dioxide?

Most of the reactions that we perform in the chemistry laboratory go essentially to completion. However, the majority of chemical reactions in our environment are in equilibrium, having a mix of reactants and products. One of the most important equilibria for our planet involves carbon dioxide. We are concerned that there is a buildup in the atmosphere of this greenhouse gas, increasing the temperature on the surface of the Earth. The amount of carbon dioxide in the atmosphere increases by about 3.4 billion tonnes a year. This is far less than the 7 billion tonnes that are released into the atmosphere each year through industrial processes and forest burning. So where do the other 3 to 4 billion tonnes of carbon dioxide go? One crucial equilibrium involves carbon dioxide dissolving in the oceans. As the atmospheric supply of carbon dioxide increases, Le Châtelier's Principle predicts that more carbon dioxide will dissolve in the oceans. It has recently been found that about 2 billion tonnes are dissolved each year. Yet that still leaves up to 2 billion tonnes unaccounted for. At this time, we are not sure where it goes. One hypothesis is that the temperate forests are recovering and absorbing the gas. Another proposal is that the increased carbon dioxide levels are leading to an increase in photosynthesis in the tropical vegetation. Many scientists are arguing that the carbon dioxide distribution is no longer an equilibrium system. To find the fate of this "missing" carbon dioxide is one of the most important tasks of today's environmental scientists.

RELATED EXPERIMENT

Experiment 33 Equilibrium and Le Châtelier's Principle

Experiment 33

Equilibrium and Le Châtelier's Principle

INTRODUCTION

A reaction in equilibrium contains both reactants and products. Many of the chemical reactions in our bodies are in equilibrium. Industrial chemists adjust the equilibrium of a reaction so that they get the maximum amount of product and the minimum amount of reactants at the end of the reaction. For example, when a company wants to produce the fertilizer, ammonia, they want only the product of the reaction:

$$N_2(g) + 3H_2(g) \rightleftharpoons 2NH_3(g)$$

In this experiment, you will investigate how to "push" the reaction to make more products. In Part I of this experiment, you will see the effect qualitatively by reacting solutions of iron(III) nitrate and potassium thiocyanate. You will determine a factor that can affect the equilibrium. In Part II, you will observe the effects that acids and bases have on a complex acid, bromothymol blue. This compound is yellow; however, when it is in its base form, it is blue.

PROBLEM

What effect does changing the concentrations of the reactants (and products) have on a reaction at equilibrium?

APPARATUS AND MATERIALS

Per pair of students:

Part I

1 96-well plate
1 toothpick
2 1-mL micro-tip pipettes
potassium thiocyanate crystals
diammonium hydrogen phosphate crystals
dropper bottles of:
    0.002 00 mol/L solution of potassium thiocyanate
    0.020 0 mol/L solution of iron(III) nitrate

Part II

1 24-well plate
1 toothpick
2 1-mL micro-tip pipettes
dropper bottles of:
bromothymol blue solution
dilute hydrochloric acid solution
dilute sodium hydroxide solution

SAFETY

Bromothymol blue is a suspected carcinogen. Hydrochloric acid, sodium hydroxide, and iron(III) nitrate are corrosive. However, in the concentrations used here, they present little risk. Wash any spills off your skin and clothing. As an extra precaution, wash your hands thoroughly after the experiment.

PART I—REACTION OF POTASSIUM THIOCYANATE SOLUTION AND IRON(III) NITRATE SOLUTION

PRELAB ASSIGNMENT

Create a table with the following headings: Procedure, Observations, and Analysis Answers. As you do each step in the Procedure, record your observations in the table. Then answer the question(s) in the Analysis that correspond to that Procedure step as in the sample table below.

PROCEDURE	OBSERVATIONS	ANALYSIS ANSWERS
1. Observe and record the colour of each solution.	*The colours are ...*	*1. The formulas are...*
2. Mix the two solutions.		*2. The new combination of ions is...*
3. Observe and record the colour of the potassium nitrate solution.	*The colour is...*	*3. (a) The ions present are...* *(b)*

PROCEDURE

1. Observe and record the colours of the solutions of potassium thiocyanate and iron(III) nitrate.

2. In a 96-well plate, use a pipette to add 15 drops of potassium thiocyanate solution to well A1. Add 1 drop of iron(III) nitrate solution to this well using a clean pipette.
3. Observe and record the colour of the potassium nitrate solution.
4. See your teacher before proceeding. Divide the coloured solution equally among wells A1, A2, B1, and B2. A1 is your reference solution.
5. Wet the toothpick with the solution in A2, and use it to pick up a tiny crystal of potassium thiocyanate. Place it in well A2 and stir. Observe and record any colour change. Clean the toothpick.
6. To well B1, add 1 drop of iron(III) nitrate solution. Observe and record any colour change.
7. Wet the toothpick with the solution in A2 and pick up a tiny crystal of diammonium hydrogen phosphate. Place it in well B1 and stir. Observe and record any colour change. If no colour change occurs, add another crystal.

Optional: Try repeating Steps 6 and 7 with the solution in well B2.

8. Discuss with your teacher what has happened before starting Part II.

ANALYSIS

NOTE: Each of the following steps relates to the step in the Procedure with the same number. For example, Step 1 in the Analysis relates to Step 1 in the Procedure.

1. Write the formulas of the ions present in the two solutions.
2. What new combination of ions resulted from this reaction?
3. (a) What ions are present in the potassium nitrate solution?
   (b) Could these ions have caused the colour that you observed? If not, what did cause this colour?
4. At this time your teacher will discuss with you what the equilibrium equation is.
5. (a) How did the addition of a crystal of potassium thiocyanate affect the concentration of the thiocyanate ion?
   (b) From your observations how has the equilibrium shifted?
6. (a) What effect does adding more iron(III) nitrate solution have on the concentration of the iron(III) ion?
   (b) How did the equilibrium shift?
7. (a) What effect does adding diammonium hydrogen phosphate have on the equilibrium?
   (b) What might the effect be of adding hydrogen phosphate ions?

PART II—EFFECT OF AN ACID AND BASE ON BROMOTHYMOL BLUE SOLUTION

PRELAB ASSIGNMENT

Write the balanced equation for this reaction using "HA" to represent the yellow form of bromothymol blue and "A^-" to represent the blue form. Draw equilibrium arrows between the reactants and products.

PROCEDURE

1. In one well of a 24-well plate, place about 1 mL of bromothymol blue solution. The well should be $\frac{1}{4}$ full.
2. Use a pipette to add the dilute hydrochloric acid solution, drop by drop while stirring the solution, until there is a colour change.
3. Use a clean pipette to then add the dilute sodium hydroxide solution, drop by drop while stirring the solution, until there is a colour change.
4. Repeat Steps 2 and 3.

ANALYSIS

1. Explain briefly why the colour changed when the acid was added in Step 2.
2. Explain briefly why the colour changed when the base was added in Step 3.
3. Did you notice an in-between colour? If you did, what caused it?

EXTENSION

1. (a) What effect do you think raising the temperature would have on the reaction in Part I? Try it with a diluted solution.
   (b) What effect would a catalyst have?
2. Read about Le Châtelier, and see if your observations agree with his principle.

Experiment 34

A Quantitative Study of Equilibrium

INTRODUCTION

In Experiment 33, you learned how to change an equilibrium reaction. But when you apply a stress, by how much will the equilibrium change? This would be a vital piece of information if you were running an industrial process.

In this experiment, you will look for patterns in the concentrations of the ions present in the following equilibrium reaction:

$$Fe^{3+}(aq) + SCN^{-}(aq) \rightleftharpoons FeSCN^{2+}(aq).$$

Remember, only the product ion has a colour but you can calculate the other two concentrations from your data.

PROBLEM

How are the concentrations of the ions in an equilibrium reaction related?

APPARATUS AND MATERIALS

Per pair of students:

1 96-well plate
5 1-mL micro-tip pipettes
dropper bottles of:
- 0.200 mol/L solution of iron(III) nitrate
- 0.002 00 mol/L solution of potassium thiocyanate
- distilled water

SAFETY

Iron(III) nitrate solution is corrosive. However, in the concentration used here, it presents little risk. Wash any spills off your skin and clothing.

PRELAB ASSIGNMENT

1. Read the Procedure. Copy and continue the sample flow chart below, to show the dilutions of the Fe^{3+} solution. Draw the chart large enough to include the calculations for the concentration of the Fe^{3+} ions in each well, as shown below.

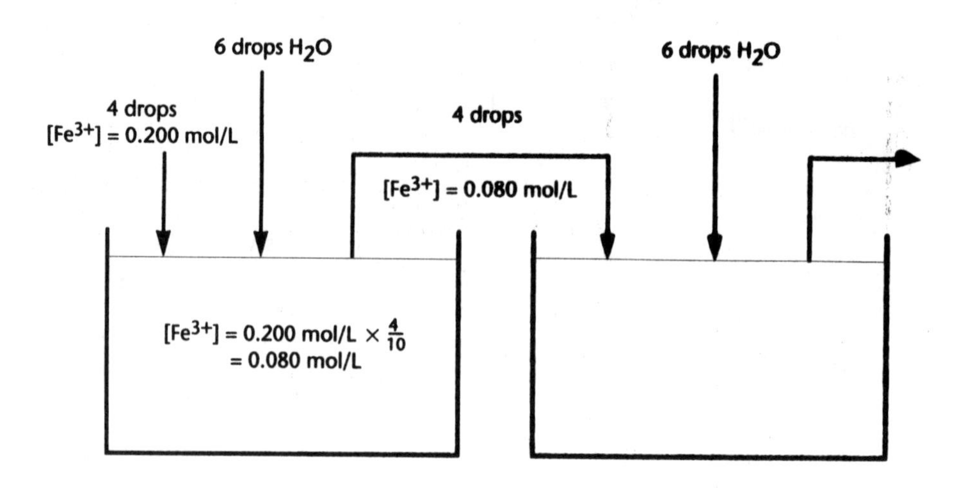

2. Make a table like the one below to record your observations.

Well	Initial		Drops of Reference	Equilibrium		
	$[Fe^{3+}(aq)]$	$[SCN^{-}(aq)]$		$[Fe^{3+}(aq)]$	$[SCN^{-}(aq)]$	$[FeSCN^{2+}(aq)]$
D3						
C3						
D4						
E3						
D2						

PROCEDURE

1. In a 96-well plate, place 5 drops of potassium thiocyanate solution in each of wells C3, D2, D3, D4, E3 as in the diagram on the next page.

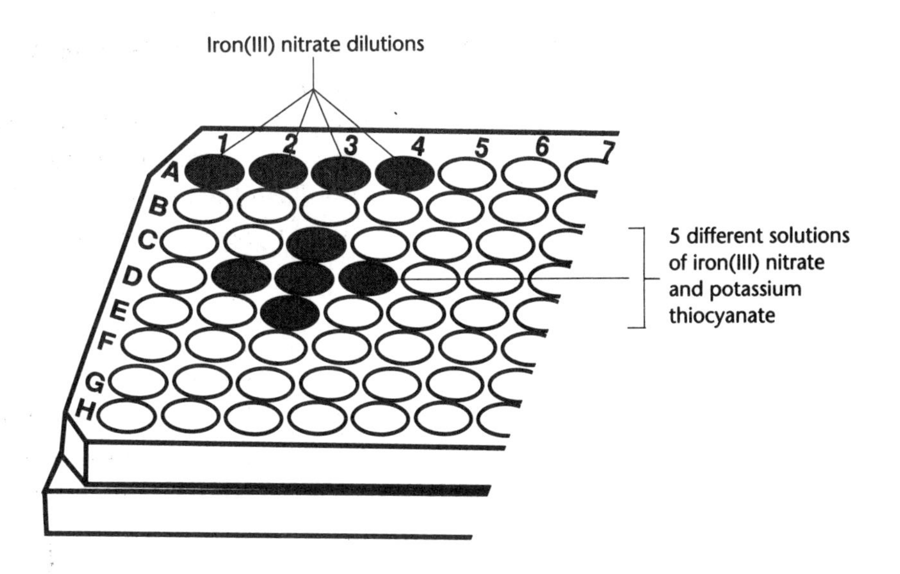

2. Add 5 drops of iron(III) nitrate solution to well D3. This is your reference solution.
3. Place 6 drops of distilled water in each of wells A1 to A4 as shown in the above diagram.
4. In well A1, add 4 drops of iron(III) nitrate solution.
5. With a clean pipette, draw up the solution from well A1 to mix it. Return it to the well. Draw up the solution a second time and try to avoid drawing up air bubbles.
6. Add 5 drops of the solution from well A1 to well C3. Add 4 drops of the solution from well A1 to well A2. Discard any remaining solution. Wash the pipette.
7. Repeat Step 5 for well A2. Add 5 drops of the solution from well A2 to well D4, and 4 drops of the solution from well A2 to well A3.
8. Repeat Step 5 for well A3. Add 5 drops of the solution from well A3 to well E3, and 4 drops of the solution from well A3 to well A4.
9. Repeat Step 5 for well A4. Add 5 drops of the solution from well A4 to well D2. The five solutions should be arranged like a five on a die with the reference solution in the centre.
10. Draw up all the reference solution in well D3 into a clean pipette. Replace the solution in well D3, drop by drop.
11. Record the number of drops in well D3 when the colour of its solution matches that of the solution in well D2 (dilution 4).
12. Continue to add drops of solution in well D3. Record the number of drops as each reference colour matches those in wells E3, D4, and C3.

ANALYSIS

1. Calculate the initial $[Fe^{3+}(aq)]$ and $[SCN^-(aq)]$ when they are mixed in each of wells D3, C3, D4, E3, and D2. Record the results in the table.

2. Calculate the equilibrium $[FeSCN^{2+}(aq)]$ by comparing the number of drops of the reference solution that had the same colour as each of the four dilutions.

3. Using the equilibrium equation, find how much $Fe^{3+}(aq)$ and $SCN^-(aq)$ were used up, and thus their equilibrium concentrations.

4. Look for a relationship among these four sets of equilibrium data. Set up a table for the four dilutions and evaluate the following expressions for each dilution. Put your answers in the table.

$$\frac{[FeSCN^{2+}(aq)]}{[Fe^{3+}(aq)] + [SCN^-(aq)]} \qquad \frac{[Fe^{3+}(aq)][Fe^{2+}SCN(aq)]}{[SCN^-(aq)]} \qquad \frac{[FeSCN^{2+}(aq)]}{[Fe^{3+}(aq)][SCN^-(aq)]}$$

5. Using the values you found for each dilution, find the ratio of the largest value to the smallest value. Which of the three expressions gives you the most consistent value?

Optional: A spreadsheet program on a computer would be helpful with your calculations.

EXTENSION

1. What is the equilibrium expression for each reaction?

(a) $PCl_5(g) \rightleftharpoons PCl_3(g) + Cl_2(g)$

(b) $N_2O_4(g) \rightleftharpoons 2NO_2(g)$

(c) $H_2CO_3(aq) \rightleftharpoons H^+(aq) + HCO_3^-(aq)$

2. In general, what is the relationship between the concentrations of the reactants and the concentrations of the products in an equilibrium reaction?

Experiment 35

Solvents and Solutes

INTRODUCTION

To understand why a compound dissolves in some solvents and not in others, you need to look at the intermolecular forces of the solute and the solvent. In this experiment, you will look for patterns in solubility and relate them to the possible solute-solvent interactions.

PROBLEM

How does solubility relate to the intermolecular forces of the solute and solvent?

APPARATUS AND MATERIALS

Per pair of students:

1 24-well plate
15 1.5-mL microtubes
3 1-mL micro-tip pipettes
1 spatula
paper towel
distilled water
ethanol
hexane
sodium chloride
sugar
naphthalene

SAFETY

Ethanol and hexane are flammable. Make sure there are no open flames in the laboratory.

PROCEDURE

1. The 24-well plate is used as a microtube rack. Place the microtubes in wells A1 to A5, B1 to B5, and C1 to C5.

2. Half-fill the pipettes. Place 0.5 mL of distilled water in each microtube in row A, 0.5 mL of ethanol in each microtube in row B, and 0.5 mL of hexane in each microtube in row C.
3. Add 5 drops of water to each microtube in wells A1 to C1. Cap and shake the microtubes.
4. Add 5 drops of ethanol to each microtube in wells A2 to C2. Cap and shake the microtubes.
5. Using the spatula, add a matchhead-sized portion of sodium chloride in each microtube in wells A3 to C3. Cap and shake the microtubes. Wipe the spatula with paper towel.
6. Add a matchhead-sized portion of sugar in each microtube in wells A4 to C4. Cap and shake the microtubes. Wipe the spatula.
7. Add a matchhead-sized portion of naphthalene in each microtube in wells A5 to C5. Cap and shake the microtubes.
8. Construct a table of solvent against solute. Record your observations of which combinations are miscible or immiscible (for liquids) and soluble or insoluble (for solids).

9. Dispose of the organic solvents as directed by your teacher.

ANALYSIS

1. (a) Draw the structural formula for each solvent molecule.
   (b) Identify which molecules are polar.
   (c) Deduce the type of intermolecular force present in each solvent.
2. (a) Identify which solutes contain ionic bonds and which contain covalent bonds.
   (b) For each covalently bonded molecule, identify the intermolecular force.
3. Explain your experimental results in terms of the comparative intermolecular forces of solute and solvent.

EXTENSION

Compare the solubilities of this series of alcohols in water: methanol, ethanol, 1-propanol, 1-butanol, 1-pentanol. Handle these alcohols with care as they are toxic and flammable. Explain your results.

ENVIRONMENTAL APPLICATION

1. Oil spills at sea can cause severe environmental problems.
   (a) Would the problem be more or less severe if oil mixed with water?
   (b) Would the problem be more or less severe if the oil did not mix, but was denser than water?
2. The organic compound DDT and a group of compounds called PCBs are considered dangerous because their molecules have low polarity. Hence, they dissolve in low polarity solvents. In mammals, the tissues containing low-polarity molecules (solvents) are in the liver and the brain. Thus, DDT and PCBs accumulate in these tissues to the level at which they can do serious physiological damage.

Insoluble Salts and the Health of Our Bones

If most minerals were soluble in water, the whole surface of this planet would be covered by water and there would be no land based life. The rocks we see around us are only there because the salts (silicates, carbonates, and phosphates) that compose them are insoluble. The insolubility of phosphates and carbonates is important to us because we belong to the vertebrates. Because our bodies have internal skeletons, we depend upon a tough, insoluble compound to provide a structural material for our bones. This compound is called calcium hydroxide phosphate, $Ca_2(OH)PO_4$, also known as apatite.

For proper bone growth and maintenance, we need to consume sufficient quantities of calcium ions and phosphate ions to exceed the solubility of apatite. Normal diets usually provide sufficient amounts of phosphate ions, but with today's preference for soft drinks over milk, many people do not have enough calcium ions in their bloodstream to form dense deposits of the apatite as bone. Thus, there is a concern about the short- and long-term health of the population. In the short term, lack of calcium ions can cause bones to break easily. Over the long term, calcium ion deficiency can increase the likelihood of developing osteoporosis in old age. In osteoporosis, bones lose calcium at a rapid rate so that the bones become extremely brittle and even a mild knock or fall can cause fractures. The solution to these problems is an adequate calcium ion intake together with vigorous exercise, which also contributes to high bone density.

RELATED EXPERIMENTS

Experiment 36 Precipitation and Equilibrium
Experiment 37 Estimating a Solubility Product Constant
Experiment 38 Calculating a Solubility Product Constant

Experiment 36

Precipitation and Equilibrium

INTRODUCTION

An application of equilibrium is in the study of solubility product constants. In previous experiments, when you reacted two solutions to form a precipitate, it was assumed that all the reacting ions in the solution were used to form the precipitate. In practice, this does not happen because the reaction will reach an equilibrium. For example, waste water from a car radiator factory would contain a high level of Zn^{2+} and Ni^{2+}. These ions can be removed by adding OH^- but an equilibrium

$$Zn(OH)_2(s) \rightleftharpoons Zn^{2+}(aq) + 2OH^-(aq)$$

can occur. Thus, some Zn^{2+} will always be in solution.

In this experiment, you will observe the reaction between a potassium chromate solution and a strontium nitrate solution.

PROBLEM

Can a precipitate be redissolved into solution?

APPARATUS AND MATERIALS

Per pair of students:

1 24-well plate
4 1-mL micro-tip pipettes
1 toothpick
dropper bottles of:
- 0.20 mol/L solution of potassium chromate
- 0.10 mol/L solution of strontium nitrate
- 0.10 mol/L solution of hydrochloric acid
- 0.10 mol/L solution of sodium hydroxide

SAFETY

Hydrochloric acid and sodium hydroxide are corrosive. The chromate ion is a suspected carcinogen. However, in the concentrations used here, they present little risk. Wash any spills off your skin and clothing. As an extra precaution, wash your hands thoroughly after the experiment.

PROCEDURE

PART I—OBSERVING THE CHROMATE/DICHROMATE EQUILIBRIUM

Here is the equation for the chromate/dichromate equilibrium.

$$2CrO_4^{2-}(aq) + 2H^+(aq) \rightleftharpoons Cr_2O_7^{2-}(aq) + H_2O(l)$$

1. Use a pipette to put enough potassium chromate solution in well F2 to cover the bottom of the well.
2. Using a pipette, add the hydrochloric acid solution drop by drop, stirring constantly with a toothpick, until a permanent colour change occurs.
3. Use a pipette to add the sodium hydroxide solution drop by drop, stirring constantly with a toothpick, until a permanent colour change occurs.
4. Transfer half the solution in well F2 to well F4.
5. Add hydrochloric acid solution drop by drop while stirring, to well F2, until the solution changes colour again. The solutions in wells F2 and F4 will be references for Part II.

PART II—CHANGING THE SOLUBILITY OF STRONTIUM CHROMATE

6. (a) Write a net ionic equation for the equilibrium reaction between potassium chromate and strontium nitrate.
   (b) Write the equilibrium constant expression.
7. Put enough potassium chromate solution in well F3 to cover the bottom of the well.
8. Add a similar amount of the strontium nitrate solution and stir. Compare this solution to the reference solutions from Part I. Record your observations.
9. Add the hydrochloric acid solution drop by drop, while stirring, until a permanent colour change occurs in the solution.

10. Add the sodium hydroxide solution drop by drop, while stirring, until a colour change occurs.

11. Repeat Steps 9 and 10. Record your observations and compare them to the reference solutions from Part I.

ANALYSIS

1. Explain what happened to the equilibrium of the solution in Part I when hydrochloric acid solution and then sodium hydroxide solution were added.
2. In Part II, what was present in the solution that was not there in Part I?
3. Explain the effect of adding
   (a) hydrochloric acid solution, and
   (b) sodium hydroxide solution to the equilibrium mixture.
4. What effect did the acid solution have on the precipitate?
5. What has happened to the equilibrium and to the concentration of the strontium ions?

EXTENSION

In the Canadian Shield, there are many lakes that have small amounts of rock dissolved in them to form an equilibrium system, like the one above. What effect would acid rain have on the concentration of ions in these lakes?

Experiment 37

Estimating a Solubility Product Constant

INTRODUCTION

In this experiment, you will produce different dilutions of the potassium chromate solution and then add each one to a fixed volume of a strontium ion solution. As you lower the concentration of the chromate ion in solution, the amount of precipitate will change. Eventually, the solution will be clear. You can then establish a range in which the solubility product constant of strontium chromate solution will lie.

PROBLEM

What is the approximate solubility product constant, K_{sp}, for the strontium chromate solution?

APPARATUS AND MATERIALS

Per pair of students:

1 96-well plate
3 1-mL micro-tip pipettes
dropper bottles of:
- 0.200 mol/L solution of potassium chromate
- 0.100 mol/L solution of strontium nitrate
- distilled water

SAFETY

The chromate ion is a suspected carcinogen. However, in the concentration used here it presents little risk. As an extra precaution, wash your hands thoroughly after the experiment.

PRELAB ASSIGNMENT

1. Write the equilibrium constant expression for strontium chromate.

2. Write the expression for the solubility product constant of strontium chromate.

3. Read the Procedure. Copy and continue the sample flow chart below to show the dilutions of the chromate solution and strontium solution. Draw the chart large enough to include the dilution calculations in each well.

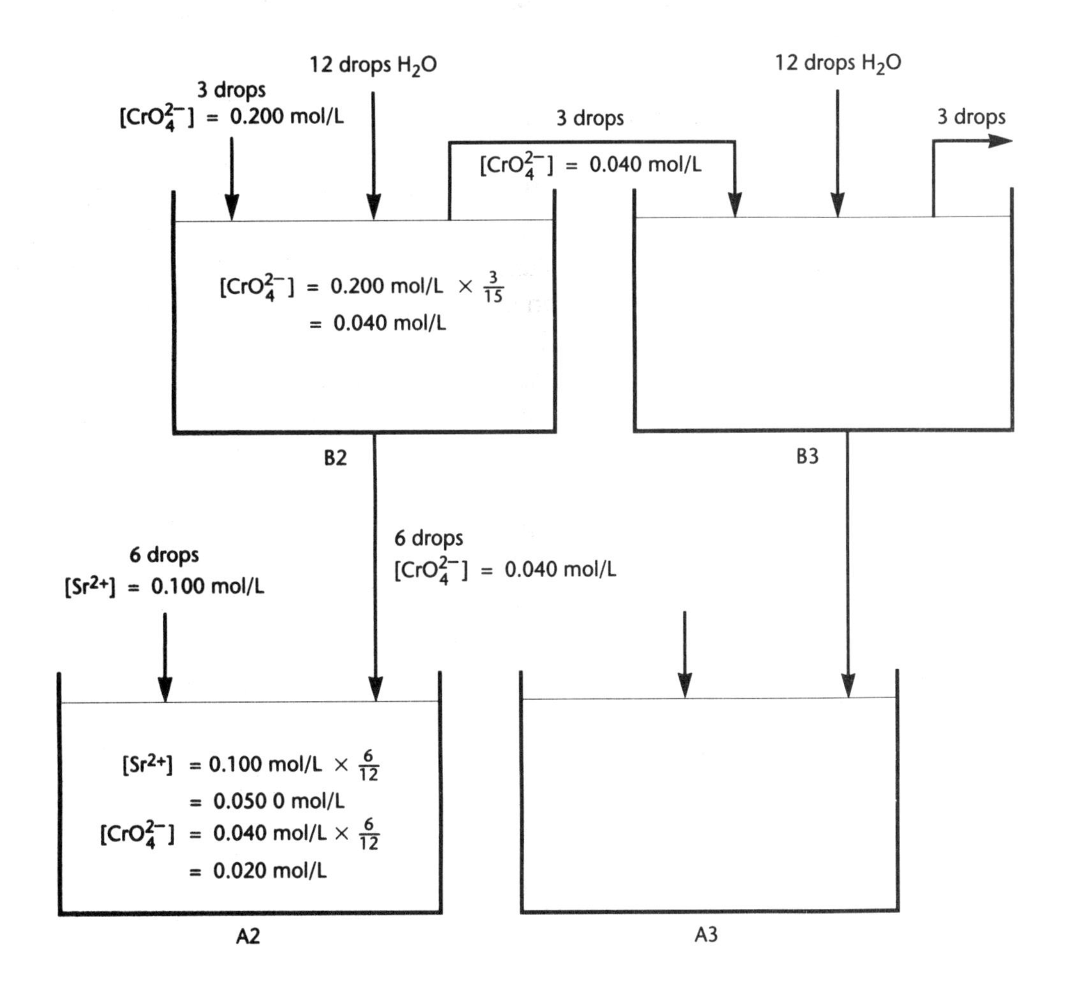

4. Copy this table.

Mixture number	Initial concentration Row A $[Sr^{2+}(aq)]$	Row B $[CrO_4^{2-}(aq)]$	Concentration when rows A and B are mixed $[Sr^{2+}(aq)]$	$[CrO_4^{2-}(aq)]$	Trial K_{sp} = $[Sr^{2+}(aq)] \times [CrO_4^{2-}(aq)]$	Precipitate formation
1						
2						
3						

PROCEDURE

1. Place 6 drops of strontium nitrate solution into each of wells A1 to A6.
2. Place 12 drops of distilled water into each of wells B2 to B6.
3. Add 6 drops of potassium chromate solution to well A1. Add 3 drops of potassium chromate solution to well B2. Dispose of the remaining solution.
4. Using a pipette, draw up the solution from well B2 to mix it. Return the solution to the well. Slowly draw up the solution from well B2 until the last drop is just inside the pipette (avoid air bubbles).
5. Add 6 drops of this solution to well A2, 3 drops to well B3, and return the remaining solution to well B2.
6. Draw up the solution from well B3, as in Step 4. Add 6 drops of this solution to well A3, 3 drops to well B4, and return the remaining solution to well B3.
7. Draw up the solution from well B4, as in Step 4. Add 6 drops of this solution to well A4, 3 drops to well B5, and return the remaining solution to well B4.
8. Draw up the solution from well B5, as in Step 4. Add 6 drops of this solution to well A5, 3 drops to well B6, and return the remaining solution to well B5.
9. Draw up the solution from well B6, as in Step 4. Add 6 drops of this solution to well A6, and return the remaining solution to well B6.
10. Hold the well plate almost parallel to a strong light to observe any precipitates formed. Record your observations in the table.

Optional: To get a narrower range for the K_{sp}, you can do an "in-between" dilution using some (3 drops) of the remaining solution in the well in row B that is opposite the last mixture in row A where you saw any precipitate. Dilute the solution in this well with an equal number of drops (three) of distilled water. Add this well-mixed solution to 6 drops of strontium nitrate solution in a well in row C, that is in same column. Observe and record any sign of a precipitate.

ANALYSIS

1. Complete the table for all the concentrations you used by inserting the concentration of each solution when the contents of wells A and B were mixed.

2. In which two adjacent wells did you see a precipitate and then no precipitate?
3. Your trial K_{sp} values then give you the range that was asked for in the Problem.
4. Compare your range for the K_{sp} of strontium chromate to the accepted value.

EXERCISES

1. As you diluted the solutions, did the mixtures across row A look clearer?
2. Have all the ions been removed from those clear wells?

EXTENSION

1. Can you suggest any way(s) this experiment could be made more exact?
2. Is dilution the solution to any pollution?
3. Would water pollution be easier to clean up at the source of the pollution or when it has reached the sewage plant? Explain.

Experiment 38

Calculating a Solubility Product Constant

INTRODUCTION

In a previous experiment, you used the colour of a solution as a measure of its equilibrium constant. In this experiment, you will calculate the solubility product constant for calcium hydroxide, which is a base. Recall that to analyse a base, you need an acid solution and an indicator.

PROBLEM

What is the solubility product constant for calcium hydroxide?

APPARATUS AND MATERIALS

Per pair of students:

1 24-well plate
3 1-mL micro-tip pipettes and labels
1 balance
1 toothpick
dropper bottles of:
- saturated solution of calcium hydroxide
- phenolphthalein
- 0.050 0 mol/L solution of hydrochloric acid

SAFETY

Calcium hydroxide and hydrochloric acid are corrosive. However, in the concentrations used here, they present little risk. Wash any spills off your skin and clothing.

PRELAB ASSIGNMENT

1. Write the equilibrium equation for the solubility of calcium hydroxide.
2. Write the solubility product constant expression for calcium hydroxide.
3. Think about why you need to analyse only the hydroxide ions.

PROCEDURE

1. Label and fill a pipette with saturated calcium hydroxide solution. Measure and record its mass.
2. Label and fill a pipette with hydrochloric acid solution. Measure and record its mass.
3. In a 24-well plate, fill a well about $\frac{1}{4}$ full with the calcium hydroxide solution.
4. Add 2 drops of phenolphthalein to this well using a pipette.
5. Titrate the calcium hydroxide solution with the hydrochloric acid solution by adding the hydrochloric acid solution drop by drop, stirring constantly.
6. As soon as a permanent colour change occurs, stop adding the hydrochloric acid solution. If you overshoot, add a few more drops of calcium hydroxide solution to the well and resume adding the hydrochloric acid solution more carefully.
7. Measure and record the mass of each pipette.
8. Repeat Steps 1 to 7 twice. Depending on the volume of solution used, you may not need to refill the pipettes.

ANALYSIS

1. For the first set of data, calculate the mass of calcium hydroxide solution used and the mass of hydrochloric acid solution used.
2. As the solutions are dilute, assume their densities are close to 1.00 g/mL. Thus, calculate the volume of each solution used.
3. Using the volume and concentration, find the amount, in moles, of hydrochloric acid used.
4. (a) Write a net ionic equation for the neutralization of the hydroxide ions by the acid.
   (b) How many moles of hydroxide ions were in your solution?

5. Find the concentration of the hydroxide ions.
6. Repeat Steps 1 to 5 for the second and third sets of data. Find the average. Use this average in the remaining calculations.
7. From the equilibrium equation for the solubility of calcium hydroxide, calculate the concentration of calcium ions.
8. Substitute the values from Steps 6 and 7 into the solubility product constant expression to find the solubility product constant for calcium hydroxide.

Optional: Use a computer spreadsheet program for these calculations.

EXERCISES

1. How does your value for the solubility product constant compare to the accepted value?
2. Your result could have been affected by a measurement you did not make. What is this measurement?
3. Can you describe a way to find the solubility product constant of a compound that does not use a titration?
4. Give three reasons why you were given calcium hydroxide to analyse rather than some other compound.

EXTENSION

1. (a) Use your answer to Exercise 3 to find the K_{sp} of calcium hydroxide.
   (b) Use a pH meter to find the pH of the original solution of calcium hydroxide. From the pH value, calculate the concentration of the hydroxide ions. Compare this value to the value you found by titration.

Experiment 39

Calculating the Molar Mass of an Acid

INTRODUCTION

Another application of an acid-base titration is to determine the molar mass of an acid. In this experiment, you will titrate tartaric acid (which is a diprotic acid) with a sodium hydroxide solution. By measuring the number of hydroxide ions used, you can calculate the molar mass of the tartaric acid. The first part of this experiment is the standardization of the sodium hydroxide solution you use. Standardization is necessary for any accurate analysis conducted in a laboratory. Even the police must standardize or calibrate their breathalyzer machine before using it. Any measuring technique should be checked regularly against a standard.

PROBLEM

What is the molar mass of tartaric acid?

APPARATUS AND MATERIALS

Per pair of students:

1 24-well plate
1 1-mL micro-tip pipette
1 toothpick
1 balance
1 spatula
0.2 g tartaric acid
0.2 g potassium hydrogen phthalate
dropper bottles of:
    0.500 mol/L solution of sodium hydroxide
    phenolphthalein
    distilled water

SAFETY

Sodium hydroxide is corrosive. However, in the concentration used here, it presents little risk. Wash any spills off your skin and clothing.

PROCEDURE

PART I—STANDARDIZATION OF SODIUM HYDROXIDE

1. Place a clean, dry 24-well plate on the balance and zero it.
2. Measure about 0.1 g of potassium hydrogen phthalate into a well. Record the mass you used.
3. Remove the well plate from the balance.
4. Add a little distilled water to cover the bottom of the well and then add 2 drops of phenolphthalein.
5. Fill a pipette with sodium hydroxide solution. Measure and record its mass.
6. Add the sodium hydroxide solution to the potassium hydrogen phthalate solution while stirring until the indicator just turns colour permanently. Measure and record the mass of the pipette.
7. Repeat Steps 1 to 6 using a second sample of potassium hydrogen phthalate.

Optional: Use a computer spreadsheet program to record and analyse your data.

PART II—TITRATION OF TARTARIC ACID

8. Repeat Steps 1 to 7 of Part I, using tartaric acid instead of potassium hydrogen phthalate.

ANALYSIS

PART I

1. (a) For the first set of data, find the moles of potassium hydrogen phthalate you used.
   (b) How many moles of sodium hydroxide solution will react with this?
2. Since you are using dilute solutions, assume that their densities are 1.00 g/mL. Use the volume of sodium hydroxide solution titrated to calculate its concentration in moles per litre.
3. Repeat Steps 1 and 2 for the second set of data. Find the average concentration. Use this value in Part II instead of the 0.500 mol/L given in the Apparatus and Materials list.

PART II

1. (a) For the first set of data, calculate the mass of sodium hydroxide solution used in the titration.

(b) Convert this mass to a volume. Calculate the moles of sodium hydroxide solution used.

2. (a) How many hydronium ions does each hydroxide ion react with?
   (b) Tartaric acid forms 2 hydronium ions from each tartaric acid molecule. How many moles of tartaric acid solution did you use?

3. Use the mass of the tartaric acid to find the mass of 1 mol of tartaric acid.

4. Repeat Steps 1 to 3 for the second set of data. Find the average mass of 1 mol of tartaric acid.

EXERCISES

1. How does your value for the molar mass of tartaric acid compare with the accepted value?

2. The simplest formula for tartaric acid is $CH(OH)CO_2H$ (or $C_2H_3O_3$). What is the molecular formula for tartaric acid?

3. Draw the structural formula for tartaric acid.

4. In which fruit is tartaric acid commonly found?

EXTENSION

1. Louis Pasteur was the first person to isolate two crystalline forms of tartaric acid.
   (a) What are they called and how are they different?
   (b) Which one predominates in nature?
   (c) In what book was this kind of "mirror-image" world used by the author Lewis Carroll?

Antacids

For our digestion enzymes to work, they need an acid environment. The hydrochloric acid in our stomachs provides this. However, a stomach may produce too much acid, in which case an antacid is needed. An antacid is simply a basic compound. Sodium hydroxide is not suitable as an antacid because the hydroxide ion is very corrosive and it is extremely soluble in water. If you swallowed it, it would severely damage your throat. However, magnesium hydroxide is extremely insoluble. As a suspension (milk of magnesia) it can be swallowed safely. The hydroxide ions are only released when the compound comes in contact with the stomach acid. Calcium carbonate is also popular as an antacid. The advertisements for calcium carbonate-containing antacids extol the benefits of combatting any calcium deficiency while settling your stomach. But a carbonate ion reacts with acid to give gaseous carbon dioxide, so you replace an acid stomach with a gassy stomach. While magnesium ions have a laxative effect, calcium ions can cause constipation. For this reason, some commercial antacids contain a mixture of the two cations to balance the effects. Even though some people are scared of chemicals, many will try these acid-base reactions to relieve their stomach problems! Thus, the best advice is to avoid the situations that lead to stomach upset.

RELATED EXPERIMENTS

Experiment 40 Determining the Ionization Constant of an Acid by Using an Indicator: Method 1

Experiment 41 Determining the Ionization Constant of an Acid by Measuring Conductivity

Experiment 42 Determining the Ionization Constant of an Acid by Using an Indicator: Method 2

Experiment 40

Determining the Ionization Constant of an Acid by Using an Indicator: Method 1

INTRODUCTION

There are many ways of determining the ionization constant of a weak acid. One method is to identify the concentration of hydronium ion in the acid. In this experiment you will compare an acetic acid solution with solutions of known hydronium ion concentration. We can make such a comparison using the colour of an indicator.

PROBLEM

What is the ionization constant, K_a, of acetic (ethanoic) acid?

APPARATUS AND MATERIALS

Per pair of students:

1 96-well plate
4 1-mL micro-tip pipettes and labels
1 toothpick
paper towel
bromophenol blue (br-bl) indicator solution
0.001 00 mol/L solution of acetic (ethanoic) acid
0.001 00 mol/L solution of hydrochloric acid
distilled water

SAFETY

Bromophenol blue is a suspected carcinogen. Acetic acid and hydrochloric acid are corrosive. However, in the concentrations used here, they present little risk. Wash any spills off your skin and clothing. As an extra precaution, wash your hands thoroughly after the experiment.

PROCEDURE

1. Fill three of the pipettes with hydrochloric acid solution (HCl), distilled water, and bromophenol blue (br-bl), and label them. In the 96-well plate, place drops of the solutions as indicated by the table below. Make sure you hold the pipettes vertically to ensure a consistent drop size.

	A1	A2	A3	A4	A5	A6	A7	A8	A9
HCl	9	8	7	6	5	4	3	2	1
water	0	1	2	3	4	5	6	7	8
br-bl	1	1	1	1	1	1	1	1	1

2. Fill a pipette with acetic acid solution. In well A12, place 9 drops of acetic acid solution and 1 drop of bromophenol blue indicator. Stir the contents of each well. (Clean the toothpick with water and a paper towel before placing it in the next solution.)
3. Match the colour (to your best judgment) of well A12 to one of the wells containing the diluted hydrochloric acid solution. Note which diluted hydrochloric acid solution matches the acetic acid solution in hydronium ion concentration.

ANALYSIS

1. Write the equilibrium equation for the ionization of acetic acid.
2. Write the expression for the ionization constant, K_a, for acetic acid.
3. For the matching well containing diluted hydrochloric acid solution, find its concentration. Use the dilution relationship, $c_1V_1 = c_2V_2$, where:

   c_1 is the initial concentration of the hydrochloric acid solution;

   V_1 is the number of drops of hydrochloric acid solution placed in the well;

   c_2 is the concentration of the diluted hydrochloric acid solution; and

   V_2 is the total number of drops of solution, indicator, and water. (Assume the indicator solution is essentially water.)

   Since we assume 100% ionization of the hydrochloric acid, the value of the hydrochloric acid concentration represents the concentration of the hydronium ion in the solution.

4. The acetic acid ionized to give the same concentration of hydronium ion as that in the matching colour of the hydrochloric acid. Since the acetic acid ionizes to give equal moles of acetate ion, this number will also be the concentration of the acetate ion.

5. Use the formula in Step 3 to calculate the concentration of the acetic acid solution after the drop of indicator solution has been added.

6. Now you know the initial concentration of acetic acid and the concentration of the equilibrium value of acetate ion. Subtract the two values to determine the concentration of acetic acid at equilibrium.

7. Substitute the values from Steps 5 and 6 in the ionization constant expression for acetic acid. Calculate the acid ionization constant value, K_a.

EXTENSION

Suppose you perform a titration of a 0.10 mol/L hydrochloric acid solution against a sodium hydroxide solution, and then repeat the experiment with a 0.10 mol/L acetic acid solution instead of the hydrochloric acid solution. Would you use more, less, or the same volume of acetic acid compared to hydrochloric acid?

Experiment 41

Determining the Ionization Constant of an Acid by Measuring Conductivity

INTRODUCTION

Ions in solution conduct electricity. In this experiment you will use a conductivity tester to test how well solutions conduct electricity. The light emitted by the tester depends upon the current flowing through the solution, and the current depends on the number of ions in solution. The weaker the light intensity, the smaller the number of ions present. In this experiment you will use the tester to find what dilution of a strong acid (hydrochloric acid) matches the same concentration of a weak acid (acetic (ethanoic) acid) in terms of the numbers of ions present.

PROBLEM

How can you use the number of ions in solution to determine the ionization constant, K_a, of acetic (ethanoic) acid?

APPARATUS AND MATERIALS

Per pair of students:

1 96-well plate
3 1-mL micro-tip pipettes and labels
1 LED conductivity tester
paper towel
1.0 mol/L solution of acetic (ethanoic) acid
1.0 mol/L solution of hydrochloric acid
distilled water

SAFETY

Acetic acid and hydrochloric acid are corrosive. Avoid contact with skin and clothing. Flush any contacted area with running water.

PROCEDURE

1. Label the first pipette "HCl," the second "HAc," and the third "H_2O."

2. In the 96-well plate, use the "HAc" pipette to place 10 drops of acetic acid solution in well A1. Always hold the pipette vertically to ensure a consistent drop size.

3. Use the "HCl" pipette to place 10 drops of hydrochloric acid solution in well B1 and 1 drop of hydrochloric acid in well B2. Dispose of the remaining acid in a waste container. Wipe the pipette tip with paper towel.

4. Use the "H_2O" pipette to add 9 drops of distilled water to well B2. Use the "HCl" pipette to mix the contents of the well thoroughly by drawing up the contents into the pipette then returning it to the well. Place 1 drop of the solution in well B3, then return the remainder to well B2. Wipe the pipette tip with paper towel.

5. Use the "H_2O" pipette to add 9 drops of water to well B3. Draw up the contents of well B3 into the "HCl" pipette. Place 1 drop of the solution in well B4, then return the remainder to well B3. Wipe the pipette tip with paper towel.

6. Use the "H_2O" pipette to add 9 drops of water to well B4. Draw up the contents of well B4 into the "HCl" pipette. Place 1 drop of the solution in well B5, then return the remainder to well B4. Wipe the pipette tip with paper towel.

7. Use the "H_2O" pipette to add 9 drops of water to well B5. Draw up the contents of well B5 into the pipette. Return the solution to well B5.

8. Insert the wires from the conductivity tester into the acetic acid solution in well A1. (Make sure the wires do not touch each other.) Note the intensity of the LED glow. Wipe the wires. Insert the wires into the hydrochloric acid solution in well B1. Compare the results.

9. You can estimate the concentration of ions in the acetic acid solution by matching the intensity of the glow in the acetic acid solution with that in one of the diluted hydrochloric acid solutions. Dip the wires into the acetic acid solution then into each of the hydrochloric acid solutions in turn until you find the closest match in intensity. Note the number of the well. Wipe the wires with paper towel between tests.

ANALYSIS

1. Calculate the concentrations of the diluted hydrochloric acid solutions in wells B2 to B5. You can use the dilution relationship, $c_1V_1 = c_2V_2$, where

   c_1 is the initial concentration of the hydrochloric acid;

   V_1 is the volume (in drops) of hydrochloric acid solution;

   c_2 is the concentration of the diluted hydrochloric acid solution; and

   V_2 is the final volume (in drops) of solution (acid plus added water).

   Since we assume 100% ionization of the hydrochloric acid, the value of each hydrochloric acid concentration represents the concentration of the hydronium ion in that solution.

2. Write the equilibrium equation for the ionization of acetic acid.

3. Write the expression for the ionization constant, K_a, for acetic acid.

4. What can you conclude about the concentration of ions in the 1.0 mol/L acetic acid solution compared to the concentration of ions in the 1.0 mol/L hydrochloric acid solution?

5. What concentration of hydronium ion in the diluted hydrochloric acid solution gave the same level of glow in the LED as the acetic acid?

6. The acetic acid ionized to give the same concentration of hydronium ions (and of acetate ions) as that in the hydrochloric acid solution of matching ion strength. Find the concentration of remaining undissociated acetic acid solution.

7. Substitute the values from steps 5 and 6 in the expression for the ionization constant. Calculate a value for the ionization constant, K_a, of acetic acid solution.

EXTENSION

1. Repeat this experiment with other concentrations of acetic acid. What do you notice about your values of the ionization constants (keeping in mind the low precision of this experiment)?

2. Repeat this experiment with other weak acids.

ENVIRONMENTAL APPLICATION

One of the reasons for purifying water is to reduce the ion content. We can measure the ion content with conductivity measurements. Purified water (distilled or de-ionized) is used in the laboratory because the ions from tap water will affect the results of experiments. Does your school laboratory use distilled or deionized water for preparing solutions? Which type of purified water can be obtained from tap water with the least amount of energy? Discuss the comparative advantages and disadvantages of the two types of purified water.

Experiment 42

Determining the Ionization Constant of an Acid by Using an Indicator: Method 2

INTRODUCTION

In science, it is important to obtain quantitative data by a number of different methods. This ensures that the reported value is accurate and precise. Another way to measure the ionization constant of a weak acid is to determine the pH of a mixture of an acid and its conjugate base using an indicator. In this experiment, you will use this method to find the ionization constant of acetic acid.

PROBLEM

What is the ionization constant, K_a, of acetic (ethanoic) acid?

APPARATUS AND MATERIALS

Per pair of students:

1 96-well plate
3 1-mL micro-tip pipettes
1 toothpick
paper towel
bromophenol blue (br-bl) indicator solution
3.0 mol/L solution of acetic (ethanoic) acid
0.50 mol/L solution of sodium acetate solution

SAFETY

Bromophenol blue is a suspected carcinogen. Acetic acid is corrosive. Avoid contact with skin and clothing. Flush any contacted area with running water. As an extra precaution, wash your hands thoroughly after the experiment.

PROCEDURE

1. Fill the pipettes with acetic acid solution (HAc), sodium acetate solution (NaAc), and bromophenol blue solution (br-bl) and label them. In the 96-well plate, place drops of solution according to the following table. Make sure you hold the pipettes vertically to ensure a consistent drop size.

	A1	A2	A3	A4	A5	A6	A7	A8	A9
HAc	9	8	7	6	5	4	3	2	1
NaAc	0	1	2	3	4	5	6	7	8
br-bl	1	1	1	1	1	1	1	1	1

2. Stir the contents of each well (clean the toothpick with paper towel between stirrings). Note the well in which the indicator is at its mid-colour between its acid and base forms. Look for the highest numbered well with the "pure" acid colour and the lowest numbered well with the "pure" base colour, then choose the transition well halfway between.

ANALYSIS

1. Write the equilibrium equation for the ionization of acetic acid.
2. Write the ionization constant expression, K_a, for the acetic acid.
3. The pH at which the mid-colour occurs for bromophenol blue is 3.85. Calculate the concentration of hydronium ion in that well.
4. For the well in which the indicator changes colour, use the dilution relationship ($c_1V_1 = c_2V_2$) to find the concentration of the acetic acid in the mixture (assume the indicator solution is mainly water). Perform a similar calculation to calculate the concentration of the acetate ion in the mixture.
5. Calculate the ionization constant, K_a, of acetic acid.

ENVIRONMENTAL APPLICATION

This experiment uses the properties of a buffer mixture. Buffering is important in many biological systems, including lake waters. Discuss why lakes are more at risk from acid rain in granite regions than in those where the bedrock is limestone.

Bases in Our Lives

We are familiar with the acids in our lives—the hydrochloric acid in our stomachs, the sulfuric acid in car batteries, and the acetic (ethanoic) acid in vinegar. Bases are less apparent. A home contains many dangerous bases, such as oven cleaners and drain openers, as well as less hazardous ones such as antacids and glass cleaners.

However, we come into daily contact with many other bases. Two of our common drugs are bases—nicotine and caffeine. It is rarely appreciated that caffeine is present in tea and many soft drinks as well as in coffee. Strong pain killers, such as morphine and codeine, are also bases. So are many of the illicit drugs, such as heroin and cocaine. In fact, cocaine is often referred to in the drug world as "free-base." All of these bases contain nitrogen atoms which behave in a similar way to ammonia. These basic groups bond to receptor sites in the brain, inhibiting pain and/or causing euphoria. Research the chemical structure of any medication that you have to take. There is a high probability that it, too, is a base.

RELATED EXPERIMENT

Experiment 43 Determining the Ionization Constant of a Base by Using an Indicator

Experiment 43

Determining the Ionization Constant of a Base by Using an Indicator

INTRODUCTION

In the three preceding experiments, the ionization constant of an acid was determined. It is equally important to find the ionization constant of a base. One way is to determine the pH of a mixture of a base and its conjugate acid, using an indicator. In this experiment, you will find the ionization constant of ammonia.

PROBLEM

What is the ionization constant, K_b, of ammonia?

APPARATUS AND MATERIALS

1 96-well plate
3 1-mL micro-tip pipettes and labels
1 toothpick
paper towel
thymol blue (th-bl) indicator solution
0.10 mol/L solution of ammonia
1.0 mol/L solution of ammonium chloride

SAFETY

Thymol blue is a suspected carcinogen. Ammonia is corrosive. However, in the concentrations used here, they present little risk. Wash any spills off your skin and clothing. As an extra precaution, wash your hands thoroughly after the experiment.

PROCEDURE

1. Fill the pipettes with ammonia solution (NH_3), ammonium chloride solution (NH_4Cl), and thymol blue solution (th-bl), and label them. In the 96-well plate, place drops of solution from the pipettes according to the following table. Make sure you hold the pipettes vertically to ensure a consistent drop size.

	A1	A2	A3	A4	A5	A6	A7	A8	A9
NH_3	9	8	7	6	5	4	3	2	1
NH_4Cl	0	1	2	3	4	5	6	7	8
th-bl	1	1	1	1	1	1	1	1	1

2. Stir the contents of each well (clean the toothpick with paper towel between stirrings). Note the well in which the indicator is at its mid-colour between its acid and base forms. Look for the highest numbered well with the "pure" base colour and the lowest numbered well with the "pure" acid colour, then choose the transition well halfway between.

ANALYSIS

1. Write the equilibrium equation for the ionization of ammonia.
2. Write the expression for the base ionization constant, K_b, for ammonia.
3. The pH at which the mid-colour occurs for thymol blue is 8.90. Calculate the concentration of hydroxide ion in that well.
4. For the well in which the indicator changes colour, use the dilution relationship ($c_1V_1 = c_2V_2$) to find the concentration of the ammonia in the mixture (assume the indicator solution is mainly water). Perform a similar calculation to calculate the concentration of the ammonium ion in the mixture.
5. Calculate the ionization constant, K_b, of ammonia.

Experiment 44

Properties of Transition Metal Ions

INTRODUCTION

Until now, almost all the reactions you have studied have used main-group metal ions, such as magnesium. Transition metal ions, such as cobalt, nickel, and copper, have some unique properties. In particular, the colours of the ions are affected by the molecules or ions surrounding them. In this experiment, you will investigate these colour changes.

PROBLEM

In what way are the transition metal ions different from the main-group metal ions?

APPARATUS AND MATERIALS

Per pair of students:

1 96-well plate
1 toothpick
paper towel
dropper bottles of:
- 0.50 mol/L solution of cobalt(II) nitrate
- 0.50 mol/L solution of nickel nitrate
- 0.50 mol/L solution of copper(II) nitrate
- 0.50 mol/L solution of magnesium sulfate
- 3.0 mol/L solution of ammonia
- 0.50 mol/L solution of sodium hydroxide

SAFETY

Ammonia and sodium hydroxide are corrosive. Avoid contact with skin and clothing. Flush any contacted area with running water.

PROCEDURE

1. In the 96-well plate, place 1 drop of the following solutions in each of the wells as indicated below. Stir the contents of each well with the toothpick and wipe the toothpick with the paper towel between stirrings. Record your observations.

Solution	Wells
cobalt(II) nitrate	A1 to A3
nickel nitrate	B1 to B3
copper(II) nitrate	C1 to C3
magnesium sulfate	D1 to D3

2. Add 1 drop of the following solutions to the wells as indicated below. Stir the contents of each well. Wipe the toothpick between stirrings. Record your observations.

Solution	Wells
ammonia	A2 to D2
sodium hydroxide	A3 to D3

ANALYSIS

1. (a) What colour was each of the transition metal ions (cobalt, nickel, and copper)? (Note: This is the colour of the transition metal ion when it is surrounded by six water molecules.)
   (b) What was the colour of the main-group metal ion (magnesium)?
2. (a) What was the colour of each transition metal ion when it was surrounded by six ammonia molecules?
   (b) Write three chemical equations to represent the replacement of water molecules around the metal ions by ammonia molecules.
   (c) What happened to the main-group metal ion when ammonia was added?
3. (a) What happened to each ion when sodium hydroxide solution was added?
   (b) Write a chemical equation to represent the each double displacement reaction that occurred.

EXTENSION

Repeat the experiment with other transition metal salts provided by your teacher.

Experiment 45

Redox Reactions of Metals and Halogens

INTRODUCTION

In chemistry there are several different building blocks or particles. The largest of these particles are ions and atoms. When studying the reactions of acids and bases, you learned that a small ion, the hydrogen ion (or proton) was the key particle in the reactions. In this experiment, you will investigate the smallest particle that can be exchanged—the electron.

In the early days of chemistry it was well known that most elements reacted with oxygen, and these reactions were called *oxidation* reactions. Metal ores are composed of compounds. To remove the less useful elements, hydrogen was first used to reduce the ores to their metals. Hence, the term *reduction* is used to describe this type of reaction.

Today, we know that both these processes involve the transfer of electrons. In this experiment, you will discover which substances gain electrons most easily and therefore reduce most easily.

PROBLEM

What metal ions and halogens are most easily reduced?

APPARATUS AND MATERIALS

Per pair of students:

1 96-well plate
1 copper wire or strip
1 lead strip
1 zinc strip
1 piece of emery paper
9 1-mL micro-tip pipettes
paper towels
dropper bottles containing 0.100 mol/L solutions of:
- copper(II) nitrate
- lead(II) nitrate
- zinc nitrate
- sodium chloride

sodium bromide
sodium iodide

dropper bottles containing aqueous solutions of:

bromine
chlorine
iodine

SAFETY

Powdered lead is very toxic. A cleaned piece of lead will be provided by your teacher. Lead(II) nitrate is also toxic and can cause birth defects if ingested by pregnant women. However, in the concentration used here, it presents little risk. As an extra precaution, wash your hands after the experiment. The bromine and iodine solutions are corrosive and can stain skin and clothing. Avoid contact with skin and clothing. Flush any contacted area with running water.

PRELAB ASSIGNMENT

1. Read the Procedure. Make a table and list the copper, lead, and zinc metal ion solutions in the first column and the names of the solid metals in the first row. Record your observations in this table.
2. Make a second table and list the three halide solutions (chloride, bromide, and iodide) in the first column and the aqueous halogen solutions in the first row. Record your observations in this table.
3. Your teacher will demonstrate the reaction of copper wire and silver nitrate solution. Use your observations to decide whether copper ions or silver ions are more reactive (more easily reduced).

PROCEDURE

PART I—REDUCTION OF METAL IONS

1. In a 96-well plate, place 3 drops of each metal ion solution in wells as indicated below.

Metal ion solution	Wells
$Cu^{2+}(aq)$	A1 to A3
$Pb^{2+}(aq)$	B1 to B3
$Zn^{2+}(aq)$	C1 to C3

2. Using emery paper, clean one end of each piece of metal except the lead. Your teacher will provide you with a clean strip of lead.
3. Add the cleaned end of the copper wire to the copper metal ion solution.
4. Wait for 30 s, remove the metal and record in your table any changes in its appearance. Wash and dry the metal.
5. Repeat Steps 2 to 4 with the other metal ion solutions.
6. Repeat Steps 2 to 5 using the lead strip and then the zinc strip.

PART II—REDUCTION OF HALOGENS

7. In another part of the 96-well plate, place 3 drops of each halide (chloride, bromide, or iodide) as indicated below, using a different pipette for each solution.

Halide solution	Wells
Cl^-*(aq)*	F1 to F3
Br^-*(aq)*	G1 to G3
I^-*(aq)*	H1 to H3

8. Using a different pipette for each solution, add 3 drops of each halogen (chlorine, bromine, or iodine) to each well as indicated below.

Halogen	Wells
Cl_2*(aq)*	F1 to H1
Br_2*(aq)*	F2 to H2
I_2*(aq)*	F3 to H3

9. Observe the wells and record the reactions in your table.

10. Dispose of the lead ion solution as instructed by your teacher.

ANALYSIS

PART I

1. (a) Which metal ion was reduced by two metals?
   (b) Which metal ion was reduced by one metal?
2. Arrange the metal ions in order of decreasing reactivity. Write the metal ion/metal half-reaction for each reaction.

PART II

3. (a) Which halogen reacted twice?
   (b) Which halogen reacted once?

4. Arrange the halogens in order of decreasing reactivity. Write the halogen/halide half-reaction for each reaction.

5. Combine the two lists of half-reactions in order of decreasing reactivity. To do this you need to know that silver ions are more easily reduced than iodine but less easily reduced than bromine. Iodine is more easily reduced than copper ions but less easily reduced than silver ions.

EXERCISES

1. Could a copper ion solution be stored safely in a zinc container? Explain.

2. Would a silver ring be safe if it was put in a chlorine bleach solution? Explain.

3. What would happen if tincture of iodine was spilled on a silver ring?

EXTENSION

Look at a larger redox chart.

1. (a) What two groups of elements are generally the most reactive?
   (b) Where are these elements found on the Periodic Table?
   (c) When were these elements discovered compared to most of the other elements?

Experiment 46

Electrochemical Cells

INTRODUCTION

In Experiment 45, you saw qualitatively which metal ion/metal combinations produced spontaneous chemical reactions.

In this experiment, you will construct various electrochemical cells. You will investigate whether different metal combinations produce different amounts of electricity. In these cells the electrons are transferred through a metal wire rather than through direct contact. This rerouting of electrons is extremely useful because the electron flow (electricity) can be used in locations far from the source of the electron flow.

PROBLEM

Which metal ion/metal pair produces the greatest voltage?

APPARATUS AND MATERIALS

Per pair of students:

1 96-well plate
1 copper strip or wire
1 lead strip
1 zinc strip
1 silver strip
4 1-mL micro-tip pipettes
1 piece of emery paper
4 strips of paper towel, 0.2 cm by 2 cm
1 multimeter tester with probes or a voltmeter with 2 electrical probes (Use a demonstration voltmeter to determine the direction of the current first.)
dropper bottles containing 0.500 mol/L solutions of:
- copper(II) nitrate
- lead(II) nitrate
- zinc nitrate
- silver nitrate
- ammonium nitrate (salt bridge solution)

SAFETY

Lead(II) nitrate is toxic and can cause birth defects if ingested by pregnant women. However, in the concentration used here, it presents little risk. As an extra precaution, wash your hands thoroughly after the experiment. Silver nitrate can stain skin and clothing.

PROCEDURE

You will make four half-cells. The copper ion/copper half-cell is the reference half-cell to which all other half-cells will be compared. The other three metals are zinc, lead, and silver. Their corresponding ionic solutions are zinc nitrate, lead(II) nitrate, and silver nitrate.

1. Clean each metal strip except the lead with emery paper. Your teacher will provide you with a clean strip of lead.
2. In a 96-well plate, place 5 drops of the copper(II) nitrate solution in well C2.
3. In three wells adjacent to well C2, place 5 drops of zinc nitrate, lead(II) nitrate, and silver nitrate solutions as shown in the diagram below.

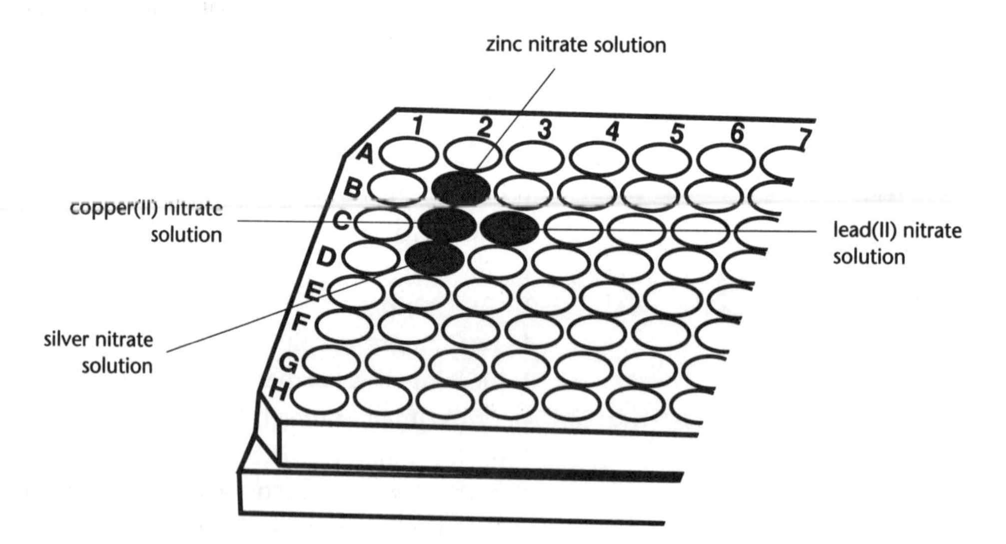

4. Place the zinc strip in the zinc nitrate solution.
5. Soak a strip of paper towel in ammonium nitrate solution. Join the copper(II) nitrate solution to the zinc nitrate solution with the paper towel strip.

6. Briefly dip the copper strip into its solution in well C2. Touch the multimeter probes to the metal strips (red probe to copper and black probe to zinc) as shown in the diagram below. Note and record only the direction of the deflection of the needle.

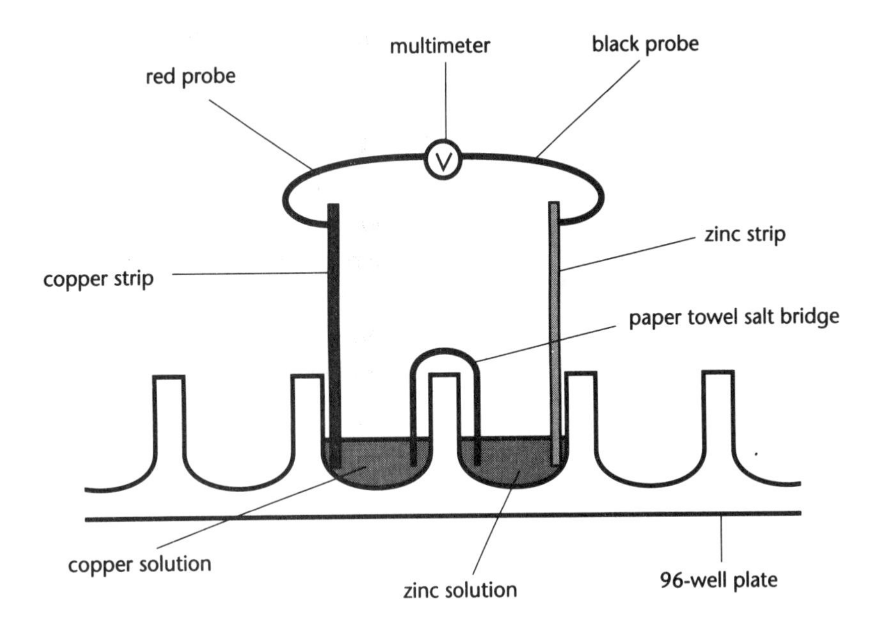

7. If the needle goes to the right, dip the copper strip into its solution again and record the maximum voltage obtained. If the needle goes to the left, reverse the probes on the metals, then record the maximum voltage obtained. Remove the paper towel strip.

8. Repeat Steps 4 to 7 for each metal strip placed in its nitrate solution.

9. Dispose of the lead ion solution according to your teacher's instructions.

ANALYSIS

1. (a) Which metal ion gains the most electrons? Which metal ion gains the least electrons?
   (b) Can you relate this to a previous experiment?

2. Write the half-reaction for each cell with its standard potential, E^0.

EXERCISES

1. (a) How would the voltage reading differ if there was more solution in each well?
   (b) What would this change in volume of solution affect?

2. Relate Exercise 1 to the different kinds of batteries (cells) you can buy: AA or C or D.

3. What would happen if the paper towel salt bridge was omitted?

EXTENSION

1. Repeat the experiment with other metals and their ion solutions to attempt to get a larger voltage.

2. Have a competition to see who can get the largest voltage, using only row A on a 96-well plate.

The Battle against Corrosion

One of the major battles of our technological society is against corrosion. Occasionally, we enjoy corrosion. The beautiful green colour of the roofs of the Parliament Buildings and the green hue of the Statue of Liberty are both caused by the oxidation (corrosion) of copper. However, we usually fight corrosion. Corrosion costs our economy billions of dollars a year. As individuals, we watch the rust spots on our cars grow like fungi. Our governments spend millions of dollars repainting bridges to inhibit rusting. It is unfortunate that iron, the cheapest and most versatile metal, should have this one flaw—easy oxidation. Chemistry predicts that we will ultimately lose the war against corrosion. The oxidation of iron is a spontaneous process, and the most fundamental statement in chemistry is that all spontaneous processes will occur, whether or not we want them to. All we can do is delay the inevitable by painting the iron to prevent moist oxygen from reaching the surface of the iron. However, as soon as the paint surface is pitted or cracked, the reaction will commence and chemistry will take its course.

RELATED EXPERIMENT

Experiment 47 Corrosion of Iron

Experiment 47

Corrosion of Iron

INTRODUCTION

One example of a redox reaction is the corrosion of iron. This is a spontaneous process that causes billions of dollars of damage each year in North America alone.

In this experiment, you will gain an understanding of this electrochemical process. You will learn how industry can overcome this problem, as well as how it can affect your car.

PROBLEM

What chemical solutions promote the corrosion of iron? What metals can protect iron from corroding?

APPARATUS AND MATERIALS

Per pair of students:

1 96-well plate
1 6-well plate
12 strips pH paper, 3 mm long
7 iron finishing nails (2.5 to 3 cm long)
14 1-mL micro-tip pipettes
1 pair of pliers
1 hammer
1 block of wood
1 4-cm copper wire
1 4-cm magnesium ribbon
zinc sheet (0.5 by 0.5 cm)
agar solution
dropper bottles containing 0.100 mol/L solutions of:
- sodium dichromate
- sodium chloride
- hydrochloric acid
- sodium hydroxide
- sodium carbonate
- nitric acid
- sodium phosphate
- urea
- sulfuric acid
- potassium hydroxide
- potassium nitrate

tap water
dropper bottles of:
phenolphthalein
0.100 mol/L potassium hexacyanoferrate solution
0.100 mol/L iron(II) sulfate solution

SAFETY

The dichromate ion is a suspected carcinogen. Hydrochloric acid, sodium hydroxide, nitric acid, sulfuric acid, and potassium hydroxide are corrosive. However, in the concentrations used here, they present little risk. Wash any spills off your skin and clothing. As an extra precaution, wash your hands thoroughly after the experiment.

PRELAB ASSIGNMENT

1. Make a table with these headings to record your observations for Part I.

Solution	Acid	Neutral	Base	Evidence of corrosion visual	Fe^{2+} Test	none

2. In the solution column, list the formulas of the eleven 0.100 mol/L solutions and tap water.

PROCEDURE

PART I—THE EFFECTS OF VARIOUS SOLUTIONS ON THE CORROSION OF IRON

1. In a 96-well plate, place 5 drops of each solution or liquid listed in your table to wells A1 to A12. Use a clean pipette for each solution.
2. Use a strip of pH paper to test each solution for acidity, neutrality, or basicity. Record your observations in your table. Remove the pH paper from the solutions.

3. Use pliers to cut each of 3 nails into 4 pieces. Place the pieces in wells A1 to A12. Leave them overnight.

4. The next day, observe each well and nail for evidence of corrosion. Record your observations in the "visual" column of the table.

5. Place 5 drops of the iron(II) sulfate solution in well B6. Add 1 drop of potassium hexacyanoferrate solution. This is a test for Fe^{2+} ions. This will be your reference well.

6. Test each well from A1 to A12 with 1 drop of potassium hexacyanoferrate solution. Compare each well to well B6. Record your observations under the "Fe^{2+} test" column in your chart.

7. Dispose of the nails and solutions as instructed by your teacher.

Optional: Record your observations on a class table. This will allow you to draw a better conclusion.

PART II—THE EFFECT OF METALS ON THE CORROSION OF IRON

8. In a 6-well plate, place a nail in one well as a control.

9. Wrap 4 cm of copper wire around a second nail and place it in a second well.

10. Wrap 4 cm of magnesium ribbon around a third nail and place it in a third well.

11. Place the zinc sheet on a block of wood and hammer a nail through it. Push the zinc to the centre of the nail. Flatten the zinc along the nail. Place it in a fourth well.

12. Cover each nail completely with the agar solution.

13. Put the cover on the well plate and draw a diagram of each nail. Leave overnight.

14. The next day, observe the nails and record your observations on the diagrams.

ANALYSIS

PART I

1. (a) What type(s) of solutions caused corrosion?
   (b) What ion is present in most of these solutions?

2. (a) What solutions do not cause corrosion?
   (b) What ion is present in most of these solutions?

3. (a) Is an iron nail oxidized or reduced? Write a half-reaction for the reaction.
   (b) Write a half-reaction for the ion from Step 1(b) undergoing the other half of the redox reaction.
4. Write the overall equation for the reaction.

PART II

1. (a) Which metals protected the iron?
   (b) Which metals corroded the iron?
2. (a) Look closely around the magnesium-wrapped nail. What seems to be produced?
   (b) Does this agree with the reaction you wrote for Step 3 in Part I?

EXERCISES

1. (a) Why can an acid corrode iron? Use a table of standard reduction potentials to explain.
   (b) Why did two metals protect the iron nail? Use the table to explain.
2. Since corrosion is a spontaneous redox reaction, how could we protect cars from corroding besides painting them?
3. Even stainless steel can be quickly corroded by acids. How can metal containers be protected from corrosion?

EXTENSION

Ships are always in water and pipelines are buried in wet ground. Find out how ships and pipelines are protected from corrosion.

Electrolysis in Industry

Many of the materials that we utilize, such as steel, cement, and nylon, do not occur naturally. We rely on chemistry to convert useless substances into useful ones. Metals, in particular, are vital to our society. Without them we would have no means of conveying an electric current—we would even find it difficult to cut food. Aluminum is one metal that is vital to a sophisticated society. For example, most aircraft are made of aluminum. This metal is so useful because of its low density and its resistance to corrosion.

How do we obtain aluminum from its ores? Unlike iron, we cannot reduce the ore to metallic aluminum simply by heating it. Instead, we have to use electrolysis. Thus we build aluminum smelters in locations where there is inexpensive electricity. For this reason, even though Canada neither possesses any of the aluminum ore (called bauxite), nor consumes a large amount of aluminum metal, it is still an important aluminum-producing nation. The pure metal is much more valuable than the ore from which it is produced. It is this "add-on" value that benefits the Canadian economy and this raises an ethical dilemma. The aluminum ores are found in some of the more poverty-stricken countries of the world. These countries mine the ore but receive little financial benefit. In addition, some environmental damage results. Should such countries be encouraged to process their own ore, raising the standard of living to above the poverty level but reducing the income that countries like Canada earn from aluminum production?

RELATED EXPERIMENTS

Experiment 48 Electrolysis of Sodium Sulfate Solution
Experiment 49 Electrolysis of Sodium Chloride Solution
Experiment 50 Electrolysis of Zinc Bromide Solution

Experiment 48

Electrolysis of Sodium Sulfate Solution

INTRODUCTION

Electricity can be used to produce chemical reactions that would not normally occur naturally. In this experiment, you will pass an electric current through a solution and identify the products of the reaction.

PROBLEM

What products are formed when sodium sulfate solution is electrolyzed?

APPARATUS AND MATERIALS

Per pair of students:

1 24-well plate
1 electrolysis apparatus (see the diagram on the next page)
1 copper electrode inserted into a 00 stopper
1 graphite electrode inserted into a 00 stopper
1 9-V battery (or other power supply)
2 wires with clips
2 toothpicks
2 00 stoppers
matches
0.10 mol/L solution of sodium sulfate containing universal indicator solution

SAFETY

If you are using a transformer plugged into an electrical outlet as your power supply, have your teacher check the wiring and avoid touching any part of the high voltage circuit.

PROCEDURE

1. In a 24-well plate, set up the apparatus as shown in the diagram below. Make sure no air bubbles are trapped in the mouth of each tube.

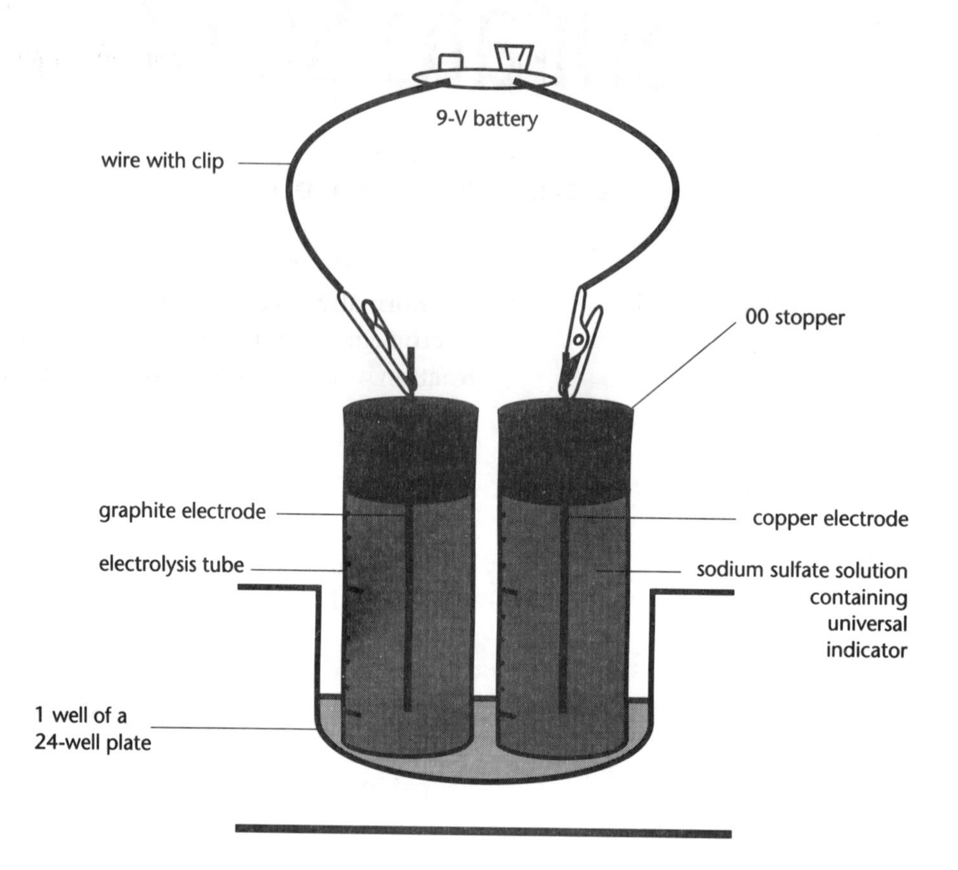

2. Using wires with clips, connect the copper electrode to the negative (black) terminal and the graphite electrode to the positive (red) terminal. Watch for any colour change in the indicator in the solution.
3. If the circuit is properly connected, bubbles should start to form on the electrodes.
4. Allow current to flow until the tube with the copper electrode is filled with gas almost to the bottom end of the electrode. Disconnect the battery.
5. Compare the volumes of gases in the tubes.
6. Lift out the tube containing the most gas, and stopper the end. Turn the tube upside down and tap the remaining solution in the tube to the other end. Hold the tube horizontally and place a burning toothpick beside the stoppered end as you remove the stopper. Record what happens.
7. Repeat Step 6 using the other tube containing gas.

ANALYSIS

1. How did the volumes of gases compare?
2. What did the change in indicator colour suggest about the ions that were being produced at each electrode?
3. Write a half-reaction for the oxidation reaction and for the reduction reaction.
4. Explain why the gases are produced in the ratios that you observed.

EXTENSION

Electrolyze other alkali metal salts provided by your teacher.

Experiment 49

Electrolysis of Sodium Chloride Solution

INTRODUCTION

The electrolysis of a sodium chloride solution has been an important commercial process. It is used to produce a variety of useful products. In this experiment, you will find out what some of these products are. In the next decade, the electrolysis of sodium chloride may no longer be used by industry, or its use may be greatly reduced. You will see why later in the Extension.

PROBLEM

What is produced when sodium chloride solution is electrolyzed?

APPARATUS AND MATERIALS

Per pair of students:

1 24-well plate
1 9-V battery (or other power supply)
2 wires with clips
1 electrolysis apparatus (see the diagram on the next page)
1 copper electrode inserted into a 00 stopper
1 graphite electrode inserted into a 00 stopper
1 toothpick
matches
starch iodide paper
red litmus paper
1 00 stopper
saturated sodium chloride solution
dropper bottle of phenolphthalein

SAFETY

In large amounts, chlorine gas is toxic. However, in this experiment only very small amounts are produced. Sniff the gas carefully. If you are using a transformer plugged into an electrical outlet as your power supply, have your teacher check the wiring and avoid touching any part of the high voltage circuit.

PROCEDURE

1. In a 24-well plate, set up the apparatus as shown in the diagram below. Make sure no air bubbles are trapped in the mouth of each tube.

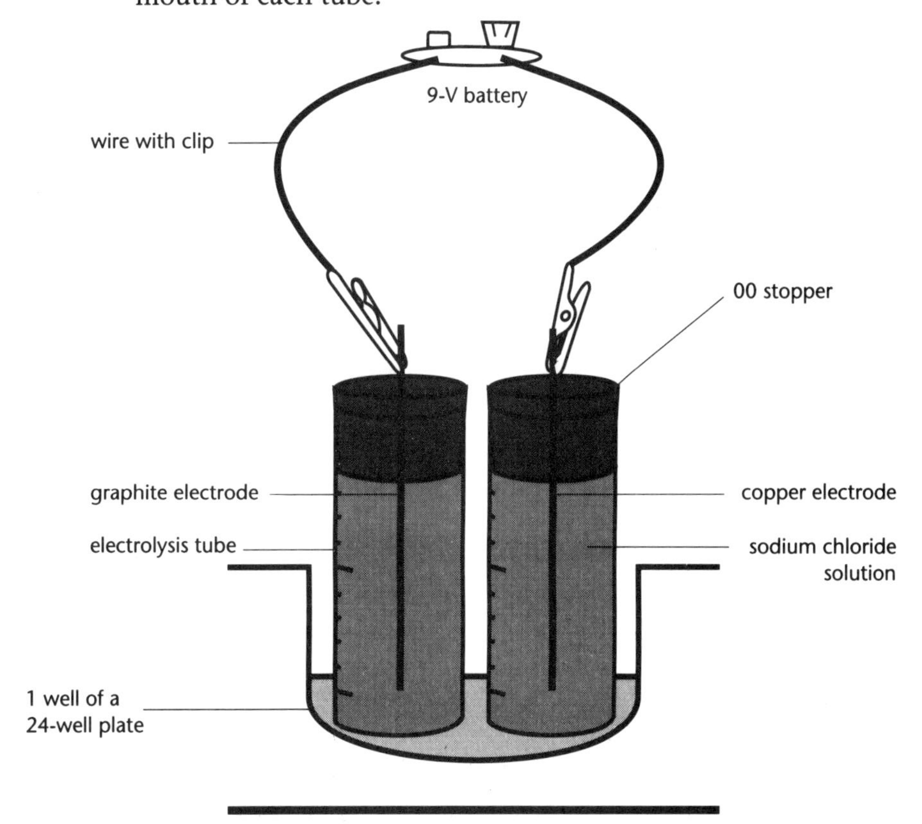

2. Using the wires with clips, connect the copper electrode to the negative (black) terminal and the graphite electrode to the positive (red) terminal.

3. If the circuit is properly connected, bubbles should start to form on the electrodes.

4. Allow current to flow until the copper electrode has its tube filled with gas to the end of the electrode. Disconnect the battery and test each electrode solution as follows.

5. For the graphite electrode—

   Add this solution to a clean well by removing the graphite electrode stopper while the open end is over a well. Carefully smell the solution, insert red litmus paper, then insert starch iodide paper. Record your observations.

6. For the copper electrode—

   Lift the tube out of the well and stopper the open end. Turn the tube upside down and tap the solution to the other end. Hold the tube horizontally and place a burning toothpick over the stoppered end as you remove the stopper. Record what happens. Hold the open tube over a clean well as you twist off the graphite electrode stopper. Insert red litmus paper into the well, and then add 1 drop of phenolphthalein. Record your observations.

ANALYSIS

1. What molecules or ions are present in your original solution?
2. Which molecules or ions would migrate to the copper electrode and which to the graphite electrode?
3. (a) From your observations, what must have reacted at the copper electrode?
   (b) Write a half-reaction for this process.
   (c) Is it an oxidation or a reduction reaction? Depending on your answer, label the copper electrode as the cathode or anode.
4. Repeat Step 3 for the graphite electrode.
5. Write the overall equation for the reaction.
6. What was left in solution?

EXTENSION

The electrolysis of sodium chloride solution is important for the production of three major chemicals.

1. What are these chemicals and what are the principal uses for each one?
2. What are some of the current problems with chlorine and its compounds?
3. (a) If the demand for chlorine goes down but the demand for sodium hydroxide remains steady, what problems would be associated with this process?
   (b) How can this be overcome?
4. Why might hydrogen peroxide replace chlorine in industrial chemical reactions?

Experiment 50

Electrolysis of Zinc Bromide Solution

INTRODUCTION

In Experiment 48, the passage of an electric current through a solution of ions caused the water to dissociate into hydrogen and oxygen. But does this always happen? In this experiment, you will electrolyze a different solution and see what is produced.

PROBLEM

What are the products in the electrolysis of zinc bromide solution?

APPARATUS AND MATERIALS

Per pair of students:

1 96-well plate
1 1-mL micro-tip pipette
1 strip of paper towel (2 cm by 0.2 cm)
1 2-cm length of copper strip (or wire)
1 2-cm carbon rod (pencil lead)
1 9-V battery (or other power supply)
2 wires with clips
dropper bottles of:
  2.0 mol/L solution of zinc bromide
  3.0 mol/L solution of hydrochloric acid
  0.10 mol/L solution of potassium iodide
mineral oil

SAFETY

Zinc bromide and hydrochloric acid are corrosive. Avoid contact with skin and clothing. Flush any contacted area with running water. If you are using a transformer plugged into an electrical outlet as your power supply, have your teacher check the wiring, and avoid touching any part of the high voltage circuit.

PROCEDURE

1. In the 96-well plate, fill two neighbouring wells with zinc bromide solution so that they are $\frac{3}{4}$ full.
2. Place the strip of paper towel across the two wells so that each end of the paper is dipping into the solution. Make sure the solution soaks through the paper.
3. Place the copper strip in one well and the carbon rod in the other well. These will act as electrodes.
4. Connect the copper strip to the negative (black) terminal and the carbon rod to the positive (red) terminal of your battery.
5. Allow the reaction to proceed until observable quantities of products have been formed at both electrodes.
6. Disconnect the battery, and remove the copper strip and carbon rod.
7. Half-fill a different well with the hydrochloric acid solution. Place the coated end of the copper strip into the solution and observe closely. Record your observations.
8. Fill the pipette $\frac{1}{4}$ full with mineral oil. Draw up some of the solution from the well in which the carbon rod had been placed until the pipette is about half full of liquid. Shake the mixture vigorously and observe. Note that the mineral oil is less dense than water. Record your observations.
9. Place some potassium iodide solution into a well.
10. From the pipette, dispose most of the lower (aqueous) layer into a waste container. Keep the mineral oil in the pipette. Draw up about some potassium iodide solution into the pipette. Shake vigorously and observe. Record your observations.

ANALYSIS

1. What did you observe at the two electrodes?
2. Based on your observations, write a half-reaction to represent the reduction reaction that you suspect occurred.
3. (a) What was the result of the test with the hydrochloric acid solution?
   (b) Did this support the half-reaction you wrote in Step 2?
   (c) Write a balanced equation to represent the reaction of the cathode product with hydrochloric acid solution.
4. Based on your observations, write a half-reaction to represent the oxidation reaction that you suspect occurred.

5. (a) What was the result of the test with the mineral oil?
   (b) Did this support the half-reaction you wrote in Step 4?
6. (a) What happened when you added the potassium iodide solution to the mineral oil?
   (b) Write a balanced equation for this reaction.

EXTENSION

Repeat the experiment using zinc iodide solution and then zinc chloride solution.

Electrochemistry: A Crucial Role in the Twenty-First Century

Of all the areas of chemistry that will revolutionize our lives, electrochemistry is one of the prime candidates. For example, we desperately need a more efficient transportation battery than the lead-acid battery. Some time in the next century we will have to rely more on non-polluting electric vehicles. Yet, without a lower-mass, longer-lasting battery, we cannot hope to reach this goal. On a larger scale, we will have to use more environmentally safe methods of electricity production such as wind-driven or wave-driven generators. Unfortunately, high winds or waves do not necessarily coincide with those time periods when we need electricity the most. So we will need batteries to store the electricity produced until it's required for use. Electrochemists are attempting to perfect fuel cells that use this surplus electrical energy to decompose water into hydrogen and oxygen by electrolysis. At times of peak demand, the two gases will be recombined to form water, releasing electrical energy.

It is in the smaller batteries that we have made the most progress. For example, the early heart pace-makers used to fail, not because of wear or mechanical defects, but more often because of the unreliable batteries of the time. By using batteries employing the newest electrochemical technology, the recipients of pace-makers no longer need to fear sudden death through battery failure. Whether you become an environmental scientist, a doctor, a nurse, or an automotive engineer, your future will be changed by electrochemists.

RELATED EXPERIMENT

Experiment 51 Constructing a Lead-Acid Battery

Experiment 51

Constructing a Lead-Acid Battery

INTRODUCTION

We can store energy by means of chemical reactions. However, in the real world, it is electrical energy that we need. In this experiment, you will convert electrical energy into chemical energy and then convert the chemical energy back into electrical energy.

PROBLEM

How can you interconvert electrical energy and chemical energy?

APPARATUS AND MATERIALS

Per pair of students:

1 24-well plate
2 2-cm lead strips
1 9-V battery (or other power supply)
2 wires with clips
voltmeter (0-3 V range)
3.0 mol/L solution of sulfuric acid

SAFETY

Lead is toxic and can cause birth defects if ingested by pregnant women. Sulfuric acid is corrosive. Avoid contact with skin and clothing. Flush any contacted area with running water. As an extra precaution, wash your hands thoroughly after the experiment. If you are using a transformer plugged into an electrical outlet as your power supply, have your teacher check the wiring, and avoid touching any part of the high voltage circuit.

PROCEDURE

1. In the 24-well plate, half-fill well A1 with the sulfuric acid solution.

2. Place the lead strips on opposite sides of well A1. Bend each lead strip over the outside edge of the well so that the strips do not come in contact with one another. These strips act as electrodes.
3. Use the wires to connect the voltmeter to the electrodes. Record the voltage (if any). Disconnect the voltmeter.
4. Use the wires with clips to connect the battery to the electrodes. Make sure the lead strips are not touching. Leave the battery connected for about 10 min. Record your observations of the electrodes.
5. Disconnect the battery and reconnect the voltmeter to the electrodes. Record the voltage.
6. If it is possible, connect a wire across the two electrodes and leave overnight. Examine the electrode surfaces the next day. Record your observations.
7. Return the lead strips to your teacher.

ANALYSIS

1. (a) Compare the voltages across the electrodes before and after a current was passed through the circuit.
   (b) How did the passage of an electric current affect the voltage?
2. Write a half-reaction for the oxidation reaction. (Hint: The lead metal is oxidized to lead(IV) oxide.)
3. Write a half-reaction for the reduction reaction. (Hint: Hydrogen ions are reduced to hydrogen gas.)
4. In this experiment, what happened to the electrical energy supplied by the battery?

ENVIRONMENTAL APPLICATION

Currently, lead-acid batteries are still the most reliable rechargeable batteries. However, the disposal of the toxic lead electrodes is a major problem. Of equal concern are the nickel-cadmium (NiCad) batteries that are popular in battery-operated home appliances and in laptop computers. The cadmium, which is used as an electrode, is a very toxic metal. Investigate what programs are available to reclaim and recycle these two harmful elements.

Experiment 52

Electroplating Copper

INTRODUCTION

In Canada, huge quantities of electricity are used in the refining of impure metals to 99.9% purity or better. On a smaller scale, the same refining process can be used to plate conducting materials with various metals like gold, silver, and copper.

In this experiment, you will plate an object with copper.

PROBLEM

At which electrode should an object be placed in order to be plated?

APPARATUS AND MATERIALS

Per pair of students:

1 24-well plate
1 copper strip
1 object to plate (a paper clip, a key)
1 strip of paper towel (1 cm by 4 cm)
1 9-V battery (or other power supply)
2 wires with clips
1 pair of tweezers
dropper bottles of:
- copper(II) sulfate plating solution
- dilute hydrochloric acid solution
- acetone
- distilled water

SAFETY

The copper(II) sulfate plating solution contains sulfuric acid. Sulfuric acid and hydrochloric acid are corrosive. However, in the concentrations used here, they present little risk. Wash any spills off your skin and clothing. Acetone is flammable. Make sure there are no open flames in the laboratory. If you are using a transformer plugged into an electrical outlet as your power supply, have your teacher check the wiring, and avoid touching any part of the high voltage circuit.

PRELAB ASSIGNMENT

Copy the diagram in the Procedure, label the flow of electrons, and predict at which electrode you need to place the object to be plated.

PROCEDURE

1. Use tweezers to hold the object to be plated. Clean it by rinsing it first in acetone, then dilute hydrochloric acid, and then distilled water.
2. In a 24-well plate, set up the apparatus as shown in the diagram below.

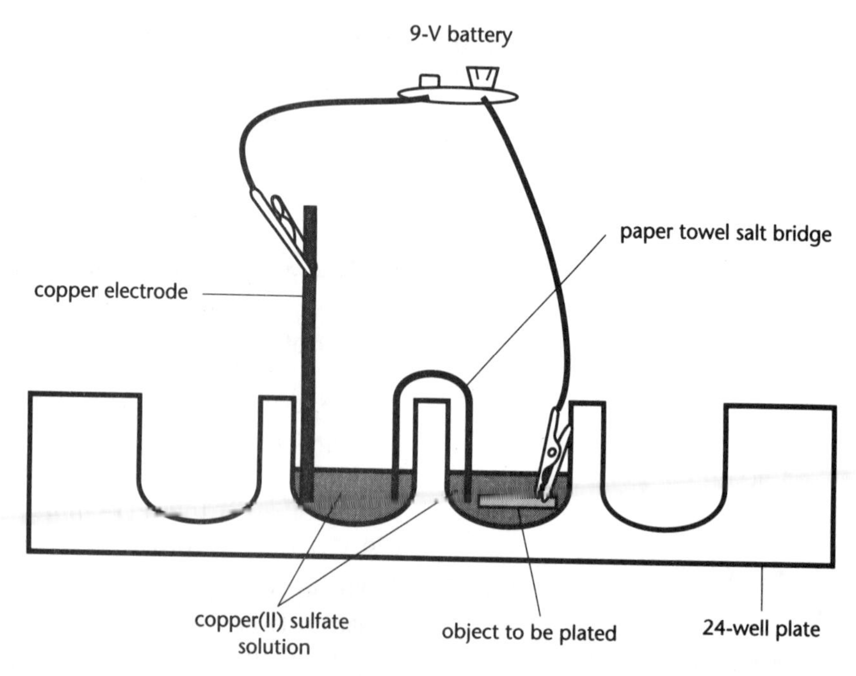

3. Connect the battery for a few minutes then check to see if any plating is occurring. If there is no plating, reverse the clips and try again. When plating is evident, disconnect the battery.
4. When you have finished, return the plating solution to your teacher.

ANALYSIS

1. (a) At which electrode did the plating occur?
   (b) Write a half-reaction for this electrode.
2. Write a half-reaction for the other electrode.
3. Write the overall equation for the reaction.

EXERCISES

1. Why was a copper electrode used?
2. If you wanted to replate a spoon with silver, what changes would you make to the apparatus and the procedure used in this experiment?

EXTENSION

The plating process is used to refine copper from 98% to 99% purity to 99.9+%. Small traces of impurities greatly affect the conductivity of copper. Common impurities in copper are iron, gold, and silver.

1. At which electrode should the impure copper be placed?
2. Write the four oxidation half-reactions for the metals present. Include the standard potentials.
3. What voltage would be needed to oxidize *just* the copper and iron into solution?
4. What voltage would be needed to keep the iron in solution and not plate out with the copper?
5. Over a period of two weeks the copper will gradually migrate from the anode to the cathode. What will happen to the gold and silver?

Experiment 53

Preparation and Properties of Acetylene

INTRODUCTION

Acetylene, more accurately called ethyne, is the simplest alkyne. Because it burns with a very hot flame, it is primarily used in welding. It is also used as a starting material for a number of industrial processes, such as the production of chlorethene and vinyl polymers. Butane, a non-alkyne, is used in cigarette lighters and as a propellant in some aerosol cans.

In this experiment, you will make some acetylene, use it in two chemical reactions, and compare its reactivity with that of butane.

PROBLEM

How do the reactivities of acetylene and butane compare?

APPARATUS AND MATERIALS

Per pair of students:

1 24-well plate
1 stopper and delivery tube
1 pair of tweezers
6 gas-collecting pipettes
1 burner
matches
1 lump of calcium carbide
dropper bottle of dilute bromine water

SAFETY

Use only a small lump of calcium carbide. Bromine is corrosive and can stain skin. Avoid contact with skin and clothing. Wash your hands after the experiment. Acetylene and butane are flammable. Make sure there are no open flames in the laboratory.

PROCEDURE

1. Fill one well of a 24-well plate about $\frac{1}{3}$ full of water.
2. Fill 4 gas-collecting pipettes with water. Stand the pipettes, open end down, in the next row of the plate. These will be used to collect gas samples.
3. Use tweezers to place a small lump of calcium carbide into the well containing water. Immediately stopper the well.
4. Place a gas-collecting pipette on top of the well, as shown in the diagram below, and let it fill with acetylene. Once it is full, remove it and place it back in the well plate. In a similar manner, completely fill a second gas-collecting pipette with acetylene. Set aside the second pipette for use in Step 12.

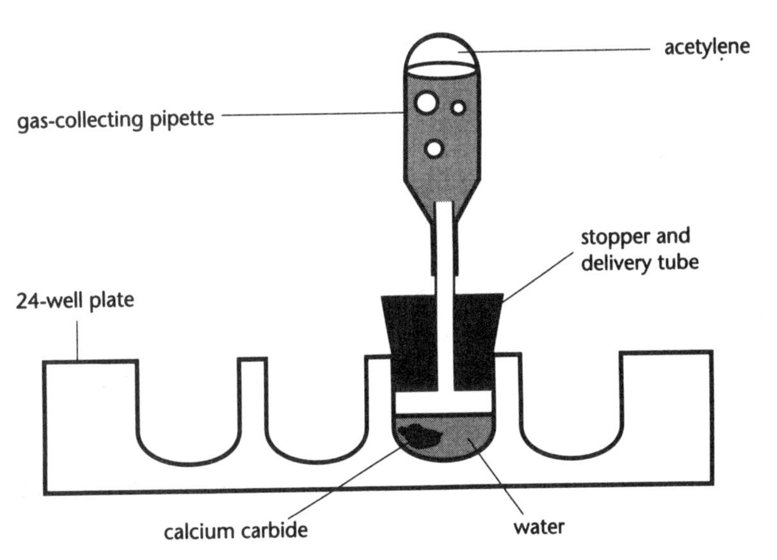

5. Place the third gas-collecting pipette on the well and half-fill it with acetylene. Place the fourth gas-collecting pipette on the well and fill it $\frac{1}{10}$ full of acetylene.

6. Place the open end of one of the completely filled pipettes about 2 cm away from a flame. Gently squirt the acetylene sideways into the flame as shown in the diagram below. Observe and record what happens.

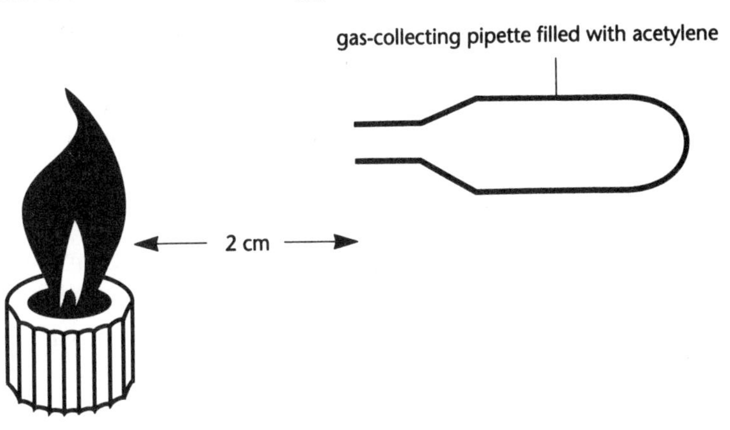

7. Squeeze out the water in the other two pipettes that contain acetylene. Release the pipettes to mix the acetylene with air.
8. Repeat Step 6 using the two other pipettes each containing different air-acetylene mixtures.
9. From your teacher, get two butane-filled gas-collecting pipettes.
10. Repeat Step 6 for a butane-filled pipette. Compare its reaction to that of the acetylene-filled pipette.
11. Fill two wells about $\frac{1}{4}$ full of dilute bromine water.
12. Slightly squeeze the acetylene-filled pipette and then draw up the bromine water from one well. Shake the pipette gently, then squirt the mixture back into the well. Record your observations.
13. Repeat Step 12, using a butane-filled pipette and the other well containing bromine water.

ANALYSIS

1. Compare how the different amounts of acetylene burned with air. Explain the differences.
2. Compare how the butane and acetylene burned.
3. How did butane and acetylene compare when mixed with bromine water?

EXERCISES

1. (a) Acetylene and butane are hydrocarbons. Which hydrocarbon burned with a sootier flame?
   (b) Draw the structures for acetylene and butane, and use the diagrams to explain your answer to part (a).
   (c) Use the diagrams in part (b) to explain the difference in reactivities of acetylene and butane.
2. (a) When a car engine or lawn mower is flooded before it is started, what does the exhaust look like?
   (b) This colour of exhaust can also be produced when some thing has happened to the air filter on the car or lawn mower. Explain.
3. Unsaturated molecules are often added to gasoline to raise its octane level. In light of this experiment, what could be a problem with the gasoline if it is left for some time in the tank?

Experiment 54

Differentiating Saturated and Unsaturated Compounds

INTRODUCTION

Alkanes (saturated compounds) are very different in their chemical behaviour from alkenes and alkynes (unsaturated compounds). In this experiment, you will conduct two tests to distinguish between saturated and unsaturated compounds.

PROBLEM

How do you distinguish between saturated and unsaturated compounds?

APPARATUS AND MATERIALS

Per pair of students:

1 24-well plate
4 1.5-mL microtubes and labels
dropper bottles of:
- cyclohexane
- cyclohexene
- bromine in dichloromethane
- potassium permanganate in acetone (propanone)

SAFETY

Cyclohexane, cyclohexene, and acetone cause dizziness if inhaled for long periods. They are also flammable. Make sure there are no open flames in the laboratory. Bromine is very corrosive and can stain skin and clothing. Handle with care. Avoid contact with skin and clothing. Flush any contacted area with running water. Dichloromethane is hazardous if inhaled. Potassium permanganate can stain skin.

PROCEDURE

1. The 24-well plate is used as a microtube rack. Place the microtubes in the plate. Place 0.5 mL of cyclohexane in each of two microtubes and 0.5 mL of cyclohexene in each of the remaining microtubes. Label the microtubes clearly.

2. Add 1 drop of bromine solution to one of the tubes containing cyclohexane. Add 1 drop of bromine solution to one of the tubes containing cyclohexene. Cap the tubes and shake them vigorously. Record your observations.
3. Add 1 drop of potassium permanganate solution to the other tube containing cyclohexane. Add 1 drop of potassium permanganate to the other tube containing cyclohexene. Cap the tubes and shake them vigorously. Record your observations.

4. Pour the solutions into an organic waste container, not in the sink!

ANALYSIS

1. (a) What did you observe when you added the bromine solution?
   (b) Write an equation to represent the reaction(s) you observed.
2. (a) What happened when you added the potassium permanganate solution?
   (b) In this test, the permanganate ion is converted to manganese(IV) oxide while the cyclohexene is converted to cyclohexane-1,2-diol. How would you describe the reaction in inorganic terms? In organic terms?

EXTENSION

React the bromine solution and the potassium permanganate solution with an unsaturated cooking oil. Do the results with this oil support the claim that the oil is unsaturated?

Optical Isomers

Experiment 55 relates to the study of structural isomers. There are several classes of isomerism, of which the most fascinating is optical isomerism. If a chemical compound contains a carbon atom to which four different groups of atoms are attached (such as CHClBrI), there are two ways in which the groups can be organized around the carbon atom, and one form will be the mirror image of the other. Hence, mirror-image versions of the same compound are referred to as optical isomers.

In conventional chemical reactions, where optical isomers are possible, they are produced in equal proportions. In the biological world, biochemical processes generally produce only one of the isomers. Each of the sugars and amino acids that we consume is a single isomer. Our bodies are unable to utilize the other isomer. If space travellers landed on a fertile planet where the opposite isomer dominated, they would starve to death!

Unfortunately, ignoring the difference in biochemical behaviour between two optical isomers has had tragic consequences. In the 1960s, a drug was discovered that prevented morning sickness in pregnant women. At the time, it was considered to be a marvellous medical breakthrough and was prescribed in many countries—until many women who took the drug gave birth to severely deformed children. Too late, the drug was withdrawn from use. It was only years later that biochemists realized that there were two optical isomers: a harmless one that prevented morning sickness and a dangerous one that caused the birth defects. With some knowledge of organic chemistry and an extra production step to separate the isomers, the catastrophe could have been avoided. But with the name thalidomide linked to one of the most dreadful episodes in modern medicine, no one today would attempt to introduce the safe isomer as a beneficial drug.

RELATED EXPERIMENT

Experiment 55 Comparison of Structural Isomers

Experiment 55

Comparison of Structural Isomers

INTRODUCTION

When the functional groups are different, we can use chemical tests to distinguish between pairs of isomers. In this experiment, you will be comparing two pairs of structural isomers: an alcohol, 1-butanol, and an ether, diethyl ether (ethoxyethane); and a ketone, acetone (propanone), and an aldehyde, propionaldehyde (propanal).

PROBLEM

How can you distinguish between structural isomers?

APPARATUS AND MATERIALS

Per pair of students:

6 1.5-mL microtubes
1 24-well plate
1 scoopula
1 set of ball-and-stick models
dropper bottles of:
- 1-butanol
- diethyl ether (ethoxyethane)
- acetone (propanone)
- propionaldehyde (propanal)
- distilled water
- 0.020 mol/L solution of potassium permanganate

copper(II) chloride

SAFETY

1-butanol, diethyl ether, acetone, and propionaldehyde are flammable. Make sure there are no open flames in the laboratory. As well, the vapours should not be inhaled deeply. Potassium permanganate solution can stain the skin.

PROCEDURE

1. The 24-well tray is used as a microtube rack. Place about 0.5 mL of 1-butanol in one microtube and about 0.5 mL of diethyl ether in another microtube. Carefully waft some of the vapour from each one towards you. Record your observations.
2. Use a scoopula to add a matchhead-sized quantity of copper(II) chloride to each microtube in Step 1. Cap each tube and shake it. Take care with the diethyl ether because pressure buildup in the tube might cause the cap to come off. Record your observations.
3. Place about 0.5 mL of 1-butanol in the third microtube and about 0.5 mL of diethyl ether in the fourth microtube. Add 2 drops of distilled water to each microtube. Cap each tube and shake it. Take care with the diethyl ether because pressure buildup in the tube might cause the cap to come off. Record your observations.
4. Place about 0.5 mL of acetone in the fifth microtube and about 0.5 mL of propionaldehyde in the sixth microtube. Carefully waft some of the vapour from each one towards you. Record your observations.
5. Add 2 drops of distilled water to each microtube in Step 4. Cap each tube and shake it. Record your observations.
6. To each mixture from Step 5, add 1 drop of potassium permanganate solution. Cap each tube and shake it. Record your observations.

7. Pour the solutions into an organic waste container, not in the sink!

ANALYSIS

1. Are the odours of 1-butanol and diethyl ether distinguishable?
2. (a) Construct a ball-and-stick model of 1-butanol and of diethyl ether.
   (b) What intermolecular forces does each molecule possess? Interpret your observations in Steps 2 and 3 of the Procedure in terms of intermolecular forces.
3. Are the odours of acetone and propionaldehyde distinguishable?
4. (a) Construct a ball-and-stick model of acetone and of propionaldehyde.
   (b) What intermolecular forces does each molecule possess? Interpret your observations in Step 5 of the Procedure.
5. (a) What is the organic product in the reaction in Step 6?
   (b) What is the name of this particular class of organic reaction?

Experiment 56

Reactions of Organic Acids and Bases

INTRODUCTION

For historical reasons, organic compounds are discussed in a completely different way from inorganic compounds. However, organic and inorganic chemistry overlap in a number of areas, one of these being acids and bases. In this experiment, you will compare organic and inorganic acids and bases.

PROBLEM

Do inorganic and organic acids and bases behave in the same way?

APPARATUS AND MATERIALS

Per pair of students:

1 24-well plate
5 1.5-mL microtubes
1 pair of tweezers
pH paper
1 scoopula
dropper bottles of:
- 3.0 mol/L solution of acetic (ethanoic) acid
- 3.0 mol/L solution of hydrochloric acid
- 3.0 mol/L solution of ammonia
- tributylamine solution
- distilled water

2 marble chips (calcium carbonate)
benzoic acid

SAFETY

Acetic acid, hydrochloric acid, ammonia, and tributylamine are corrosive. Avoid contact with skin and clothing. Flush any contacted area with running water.

PROCEDURE

1. The 24-well plate is used as a microtube rack. Place about 0.5 mL of acetic acid solution in one microtube and about 0.5 mL of hydrochloric acid solution in another microtube. Test each solution with pH paper. Record your observations.
2. Using tweezers, add one marble chip (calcium carbonate) to each solution. Record your observations.
3. Place about 0.5 mL of tributylamine solution in a clean microtube and about 0.5 mL of ammonia solution in another clean microtube. Test each solution with pH paper. Record your observations.
4. In another clean microtube, place about 0.5 mL of distilled water. Using a scoopula, add about a matchhead-sized quantity of benzoic acid to this microtube and to each microtube containing the tributylamine solution and the ammonia solution. Cap each microtube and shake it vigorously. Record your observations.

ANALYSIS

1. (a) What evidence did you obtain that acetic acid behaves similarly to a simple inorganic acid?
   (b) Write a balanced chemical equation for each chemical reaction that occurred in Step 2 of the Procedure.
2. (a) What evidence did you obtain that tributylamine behaves like an inorganic base?
   (b) Write a balanced chemical equation for each chemical reaction that occurred in Step 4 of the Procedure.

ENVIRONMENTAL APPLICATION

Even though we blame pollution for the excessive acid levels in rivers and lakes, we must take into account the "natural" contribution to water acidity. This contribution derives partially from the organic (humic) acids that are produced by the decay of vegetation. It is the humic acids that cause water from bogs to be particularly acidic.

EXTENSION

If there is a bog near your school, compare the pH of the water in the bog to that of water samples from other sources.

Experiment 57

Preparation of Esters

INTRODUCTION

Esters are useful organic compounds that are made from acids and alcohols. They occur naturally in fruits and provide the pleasant flavours associated with apples, oranges, and bananas. Minute amounts of esters are used by some insects to attract mates.

In this experiment, you will perform esterification and, by identifying the odour of the ester, identify the acid and the alcohol.

PROBLEM

What acids and alcohols were used to produce each ester?

APPARATUS AND MATERIALS

Per pair of students:

5 modified jumbo pipettes and labels
9 thin-stem pipettes
5 cut-off graduated pipettes (used as stoppers)
1 spatula
1 hot plate
1 beaker
1 thermometer
1 Styrofoam cup
solid acid samples labelled D and E
dropper bottles of:
- acid solutions labelled A to C
- alcohol solutions labelled F to J
- 9.0 mol/L solution of sulfuric acid

SAFETY

Solutions A to C and sulfuric acid are very corrosive. Handle with care. Avoid contact with skin and clothing. Flush any contacted area with running water. Avoid inhaling the dust of the solid acid samples D and E.

OPTIONAL PRELAB ASSIGNMENT

If you have access to a computer and a relevant program, you can simulate these reactions before conducting the experiment.

PROCEDURE

1. Boil about 75 mL of tap water and pour it into a Styrofoam cup.
2. Use a thin-stem pipette to add 3 drops of each acid solution into 3 modified jumbo pipettes as in shown in the diagram below.

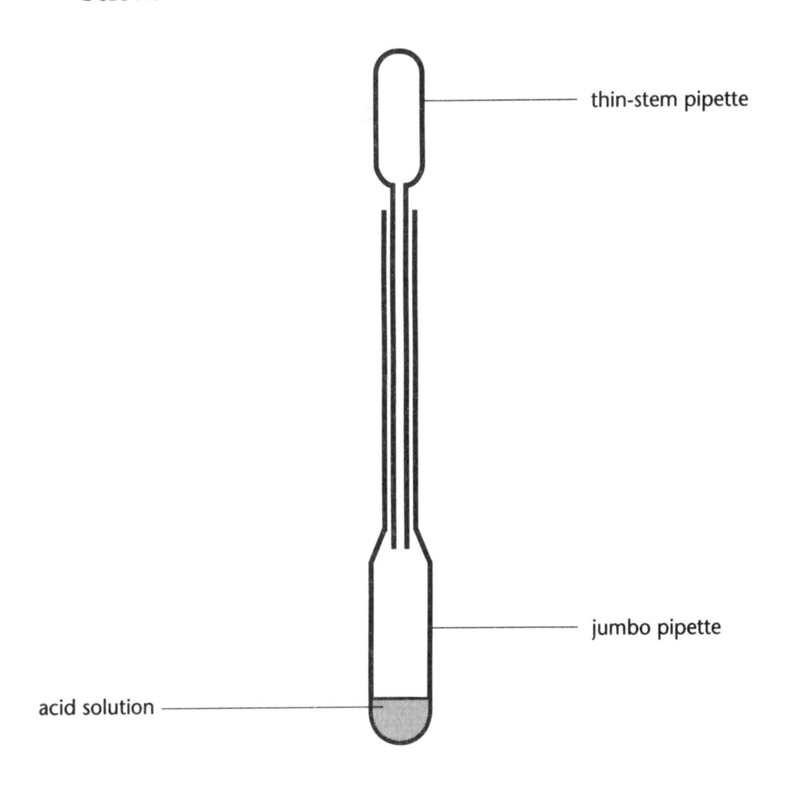

For each acid in solid form, use a spatula to insert a match head-sized quantity of acid into a modified jumbo pipette.

3. Use a thin-stem pipette to add 10 drops of each alcohol solution to the appropriate acid, as shown in the chart below.

Note: Use the acids and alcohols in the following combinations:

Acid	Alcohol
A	F
B	G
C	H
D	I
E	J

4. In the fumehood, use a thin-stem pipette to add 1 drop of 9.0 mol/L sulfuric acid solution to each jumbo pipette. Be careful.

5. Label each pipette with its contents. Put a cut-off graduated pipette in the end of each jumbo pipette. Fill the water-bath with the heated water. Put the thermometer in the water. When the temperature reaches 80°C, put the pipettes in the bath. Leave the pipettes for 10 minutes to allow the temperature of their contents to rise to between 70°C and 80°C.

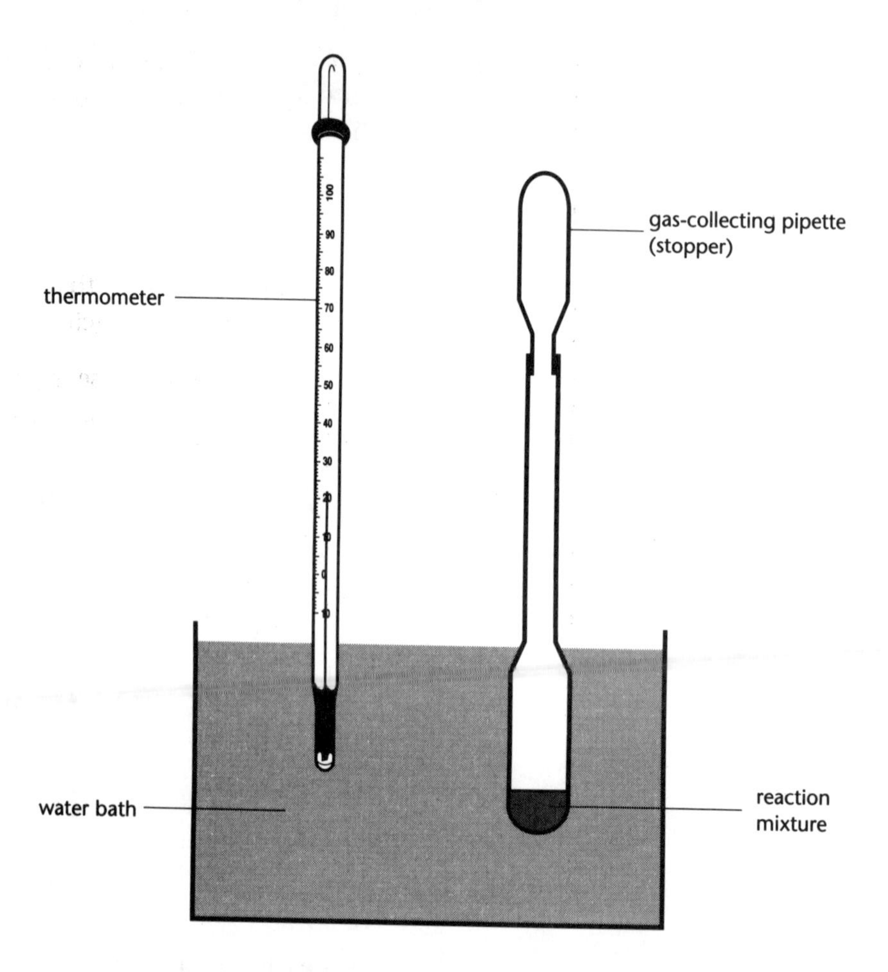

6. Remove the cut-off graduated pipette from one of the samples. Cautiously smell the odour by squirting the fumes out into the air and letting them waft towards your nose. Note your observations.

7. Repeat Step 6 for each of the other jumbo pipettes.

8. If you have a problem smelling the odours, take the pipettes outside the laboratory to smell them. There may be too many competing odours in the laboratory for you to identify accurately your own samples.

9. Use the following information to identify the acid and alcohol which produced each ester.

Odour	Compound
peppermint	ethyl benzoate
wintergreen	methyl salicylate
plums	amyl formate
bananas	isoamyl acetate
pineapple	butyl butyrate
nail polish	ethyl acetate
oranges	octyl acetate

10. Dispose of the esters and sulfuric acid solution according to your teacher's instructions.

EXERCISES

1. Write an equation for the production of each ester you identified. Name the acid and the alcohol.
2. (a) How fast were these reactions? Explain your answer.
   (b) Why was sulfuric acid used?

Chemophobia and Drugs

Chemophobia is the fear of chemicals and of chemistry. By this time in the course, we hope that you do not suffer from it! Chemophobics divide the world into "natural" and "chemical" materials. Yet most people do not realize there is no division between the two. For example, if you shop for vitamin tablets, such as vitamin C, there are often two types on the drugstore shelf. One of these is labelled "vitamin C" while the other is labelled "vitamin C—natural source"—that is, the vitamin C has been extracted from fruit rather than synthesized in a biochemical laboratory. The "natural source" vitamin C is usually at least twice the price of the regular vitamin C. Yet the Law of Constant composition tells us that vitamin C will have the same chemical composition—$C_6H_8O_6$—regardless of its source! So the manufacturers of the "natural source" products profit from people's ignorance of chemistry.

More amazingly, chemophobics often overlook their dependence on drugs, most of which come from "natural" sources. Much of the world's population is addicted to the drug caffeine, found in tea, coffee, and some soft drinks. The naturally occurring drugs, nicotine and alcohol, are even more addictive and poisonous. The pain-killer aspirin, a derivative of naturally occurring salicylic acid, is another common drug consumed by millions of people every day.

RELATED EXPERIMENT

Experiment 58 Aspirin Purity

Experiment 58

Aspirin Purity

INTRODUCTION

One of the most common drugs used today is aspirin. It provides some relief from discomfort and fever. Although aspirin is not produced commercially using the reaction equation below, it does show that aspirin is an ester.

$$\text{C}_6\text{H}_4(\text{CO}_2\text{H})(\text{OH}) + CH_3CO_2H \rightleftharpoons \text{C}_6\text{H}_4(\text{CO}_2\text{H})(\text{OCOH}_3) + H_2O$$

salicylic acid (ring with CO_2H) — aspirin (ring with CO_2H and $OCOH_3$)

The active part of aspirin is the salicylic acid which is not readily soluble in water and can irritate the lining of the stomach. Thus, in the last century, acetylsalicylic acid (aspirin) was produced to try to reduce this problem. The above equation represents an equilibrium reaction; consider what problem could arise if aspirin is stored for a long time, especially in your bathroom.

In this experiment, you will calculate the amounts of salicylic acid in samples of aspirin and investigate how these amounts are affected by the age of the aspirins.

PROBLEM

How much salicylic acid is present in various samples of aspirin?

APPARATUS AND MATERIALS

Per pair of students:

1 96-well plate
1 24-well plate
4 1-mL micro-tip pipettes
1 jumbo pipette
1 beaker
1 balance
1 mortar and pestle
1 10-mL graduated cylinder
various aspirin tablets

dropper bottles of:
- 50% ethanol solution
- 0.10 mol/L solution of iron(III) nitrate
- standard salicylic acid solution
- distilled water

SAFETY

Ethanol is flammable. Make sure there are no open flames in the laboratory.

PROCEDURE

PART I—PREPARING DILUTIONS OF THE STANDARD SALICYLIC ACID SOLUTION

1. In a 96-well plate, use micro-tip pipettes to add the standard salicylic acid solution and distilled water to wells A1 to A10 according to the following dilutions:

	A1	A2	A3	A4	A5	A6	A7	A8	A9	A10
acid	1	2	3	4	5	6	7	8	9	10
water	9	8	7	6	5	4	3	2	1	0

2. Add 1 drop of the iron(III) nitrate solution to each well from A1 to A10. Notice that the colour gradually gets stronger from well A1 to well A10. Save these reference solutions for later comparison.
3. You may need to dilute the standard salicylic acid solution. If so, use a 24-well plate and micro-tip pipette or a graduated jumbo pipette.

PART II—ANALYSIS OF ASPIRIN SAMPLES

4. Measure and record the mass of an aspirin tablet.
5. Crush the tablet and then place it in a beaker. Dissolve the tablet in 10.0 mL of 50% ethanol solution. Some of the tablet will not dissolve so allow some time for it to settle.
6. Using a micro-tip pipette, place 10 drops of the aspirin solution in well B5. Add 1 drop of iron(III) nitrate solution.
7. Compare the colour of the solution in well B5 with the reference solutions. If there is no purple colour then there is no free salicylic acid present.

8. If your aspirin solution is too dark to compare to the reference solutions, use a micro-tip pipette to add more 50% ethanol solution to make a more dilute solution that matches one of the reference solutions. Record any dilutions you did.

9. Repeat Steps 4 to 8 for other aspirin samples.

ANALYSIS

1. Calculate the mass of salicylic acid per 100 mL of solvent in each of wells A1 to A10.

2. Which well best matches the aspirin sample? Therefore, what is the concentration of the salicylic acid in that aspirin sample?

3. From the aspirin solution you made, and any dilution of it, calculate the total volume of your whole aspirin solution.

4. From Steps 2 and 3, find the total mass of salicylic acid that was in the aspirin tablet.

5. Calculate the percent mass of salicylic acid in the aspirin tablet.

6. Repeat Steps 2 to 5 for other aspirin samples.

EXERCISES

1. Look at the equation for the formation of aspirin. Since it is an equilibrium reaction, what might drive the reaction back to the reactants?

2. (a) Where would you normally store aspirin at home?
   (b) Is this a good choice in light of your answer to Exercise 1?

3. Why does a bottle of aspirin tablets have an expiry date?

EXTENSION

1. Collect a variety of old and new aspirin tablets. Store them under different conditions. Analyse the tablets at a later date and compare your results with those from this experiment.

2. Dissolve an aspirin tablet in water and study the rate of formation of salicylic acid over time.

Text References

Experiment	Chemistry A FIRST COURSE	Chemistry A SECOND COURSE	Addison-Wesley Chemistry
1. Chromatography of an Ink	1		16
2. Chemical Changes	1		1
3. Identification of Household Substances			
4. Changes in Mass during Chemical Reactions	10		1
5. Law of Definite Proportions	7		5
6. Flame Spectra	3		11
7. Boyle's Law and the Properties of Gases	11		10
8. Charles' Law and the Properties of Gases	11		10
9. Molar Volume of a Gas	11		6
10. Molar Mass of a Gas	11		6 and 9
11. Single Displacement Reactions	8		7
12. Double Displacement Reactions	8		7
13. Patterns in the Properties of Compounds	4 and 6		5 and 12
14. Stoichiometry of a Reaction Producing a Solid	10		8
15. Stoichiometry of a Reaction Producing a Gas	10		8 and 18
16. Production of Hydrogen and Oxygen	9		23
17. Percentage Yield and Purity	10		8
18. Classifying Ionic Solutions	12		15 and 16
19. Conductivity of Solutions	12		15 and 16
20. Solubilities of Salts	8		16 and 19
21. Solubility Curves	12		16
22. Identifying Colourless Solutions	12		18
23. Analysis of Vinegar	13		19
24. Volumetric Acid-Base Titration	13		19
25. Vitamin C Analysis	13		19
26. Actual and Theoretical Yields	13		7, 8, 10, and 18
27. Heats of Reaction		5	27
28. Heat of Combustion of Magnesium		5	27
29. Effect of Surface Area on Rate of Reaction		6	17
30. Effects of Concentration and Temperature on the Rate of Reaction		6	17

Experiment	Chemistry A FIRST COURSE	Chemistry A SECOND COURSE	Addison-Wesley Chemistry
31. Effect of a Catalyst on the Rate of Reaction		6	17
32. Rate Law for a Reaction		6	17
33. Equilibrium and Le Châtelier's Principle		7	17
34. A Quantitative Study of Equilibrium		7	17
35. Solvents and Solutes		8	15
36. Precipitation and Equilibrium		8	19
37. Estimating a Solubility Product Constant		8	19
38. Calculating a Solubility Product Constant		1 and 8	19
39. Calculating the Molar Mass of an Acid		10	19
40. Determining the Ionization Constant of an Acid Using an Indicator: Method 1		10	19
41. Determining the Ionization Constant of an Acid by Measuring Conductivity		10	19
42. Determining the Ionization Constant of an Acid Using an Indicator: Method 2		10	19
43. Determining the Ionization Constant of a Base by Using an Indicator		10	19
44. Properties of Transition Metal Ions		8 and 10	22
45. Redox Reactions of Metals and Halogens		11	20
46. Electrochemical Cells		12	21
47. Corrosion of Iron		12	20
48. Electrolysis of Sodium Sulfate Solution		12	21
49. Electrolysis of Sodium Chloride Solution		12	21
50. Electrolysis of Zinc Bromide Solution		12	21
51. Constructing a Lead-Acid Battery		12	21
52. Electroplating Copper		12	21
53. Preparation and Properties of Acetylene		14	25
54. Differentiating Saturated and Unsaturated Compounds		14	25
55. Comparison of Structural Isomers		15	25
56. Reactions of Organic Acids and Bases		15	26
57. Preparation of Esters		15	26
58. Aspirin Purity		15	26

Periodic Table of the Elements

Key: State — Atomic number — Element symbol — Name — Average atomic mass — Electrons in each energy level

Example: (S) 11 Na Sodium 22.990 — 2, 8, 1

State: (S) Solid; (L) Liquid; (G) Gas; (N) Not found in nature

Metals / **Nonmetals**

1† 1A	2 2A	3 3B	4 4B	5 5B	6 6B	7 7B	8 8B	9 8B	10 8B	11 1B	12 2B	13 3A	14 4A	15 5A	16 6A	17 7A	18 0
(G) 1 H Hydrogen 1.0079 (1)																	(G) 2 He Helium 4.0026 (2)
(S) 3 Li Lithium 6.941 (2, 1)	(S) 4 Be Beryllium 9.0122 (2, 2)											(S) 5 B Boron 10.81 (2, 3)	(S) 6 C Carbon 12.011 (2, 4)	(G) 7 N Nitrogen 14.007 (2, 5)	(G) 8 O Oxygen 15.999 (2, 6)	(G) 9 F Fluorine 18.998 (2, 7)	(G) 10 Ne Neon 20.179 (2, 8)
(S) 11 Na Sodium 22.990 (2, 8, 1)	(S) 12 Mg Magnesium 24.305 (2, 8, 2)											(S) 13 Al Aluminum 26.982 (2, 8, 3)	(S) 14 Si Silicon 28.086 (2, 8, 4)	(S) 15 P Phosphorus 30.974 (2, 8, 5)	(S) 16 S Sulfur 32.06 (2, 8, 6)	(G) 17 Cl Chlorine 35.453 (2, 8, 7)	(G) 18 Ar Argon 39.948 (2, 8, 8)
(S) 19 K Potassium 39.098 (2, 8, 8, 1)	(S) 20 Ca Calcium 40.08 (2, 8, 8, 2)	(S) 21 Sc Scandium 44.956 (2, 8, 9, 2)	(S) 22 Ti Titanium 47.90 (2, 8, 10, 2)	(S) 23 V Vanadium 50.941 (2, 8, 11, 2)	(S) 24 Cr Chromium 51.996 (2, 8, 13, 1)	(S) 25 Mn Manganese 54.938 (2, 8, 13, 2)	(S) 26 Fe Iron 55.847 (2, 8, 14, 2)	(S) 27 Co Cobalt 58.933 (2, 8, 15, 2)	(S) 28 Ni Nickel 58.71 (2, 8, 16, 2)	(S) 29 Cu Copper 63.546 (2, 8, 18, 1)	(S) 30 Zn Zinc 65.38 (2, 8, 18, 2)	(S) 31 Ga Gallium 69.72 (2, 8, 18, 3)	(S) 32 Ge Germanium 72.59 (2, 8, 18, 4)	(S) 33 As Arsenic 74.922 (2, 8, 18, 5)	(S) 34 Se Selenium 78.96 (2, 8, 18, 6)	(L) 35 Br Bromine 79.904 (2, 8, 18, 7)	(G) 36 Kr Krypton 83.80 (2, 8, 18, 8)
(S) 37 Rb Rubidium 85.468 (2, 8, 18, 8, 1)	(S) 38 Sr Strontium 87.62 (2, 8, 18, 8, 2)	(S) 39 Y Yttrium 88.906 (2, 8, 18, 9, 2)	(S) 40 Zr Zirconium 91.22 (2, 8, 18, 10, 2)	(S) 41 Nb Niobium 92.906 (2, 8, 18, 12, 1)	(S) 42 Mo Molybdenum 95.94 (2, 8, 18, 13, 1)	(N) 43 Tc Technetium (97) (2, 8, 18, 14, 1)	(S) 44 Ru Ruthenium 101.07 (2, 8, 18, 15, 1)	(S) 45 Rh Rhodium 102.91 (2, 8, 18, 16, 1)	(S) 46 Pd Palladium 106.4 (2, 8, 18, 18)	(S) 47 Ag Silver 107.87 (2, 8, 18, 18, 1)	(S) 48 Cd Cadmium 112.41 (2, 8, 18, 18, 2)	(S) 49 In Indium 114.82 (2, 8, 18, 18, 3)	(S) 50 Sn Tin 118.69 (2, 8, 18, 18, 4)	(S) 51 Sb Antimony 121.75 (2, 8, 18, 18, 5)	(S) 52 Te Tellurium 127.60 (2, 8, 18, 18, 6)	(S) 53 I Iodine 126.90 (2, 8, 18, 18, 7)	(G) 54 Xe Xenon 131.30 (2, 8, 18, 18, 8)
(S) 55 Cs Cesium 132.91 (2, 8, 18, 18, 8, 1)	(S) 56 Ba Barium 137.33 (2, 8, 18, 18, 8, 2)	(S) 71 Lu Lutetium 174.97 (2, 8, 18, 32, 9, 2)	(S) 72 Hf Hafnium 178.49 (2, 8, 18, 32, 10, 2)	(S) 73 Ta Tantalum 180.95 (2, 8, 18, 32, 11, 2)	(S) 74 W Tungsten 183.85 (2, 8, 18, 32, 12, 2)	(S) 75 Re Rhenium 186.21 (2, 8, 18, 32, 13, 2)	(S) 76 Os Osmium 190.2 (2, 8, 18, 32, 14, 2)	(S) 77 Ir Iridium 192.22 (2, 8, 18, 32, 15, 2)	(S) 78 Pt Platinum 195.09 (2, 8, 18, 32, 17, 1)	(S) 79 Au Gold 196.97 (2, 8, 18, 32, 18, 1)	(L) 80 Hg Mercury 200.59 (2, 8, 18, 32, 18, 2)	(S) 81 Tl Thallium 204.37 (2, 8, 18, 32, 18, 3)	(S) 82 Pb Lead 207.2 (2, 8, 18, 32, 18, 4)	(S) 83 Bi Bismuth 208.98 (2, 8, 18, 32, 18, 5)	(S) 84 Po Polonium (209) (2, 8, 18, 32, 18, 6)	(S) 85 At Astatine (210) (2, 8, 18, 32, 18, 7)	(G) 86 Rn Radon (222) (2, 8, 18, 32, 18, 8)
(S) 87 Fr Francium (223) (2, 8, 18, 32, 18, 8, 1)	(S) 88 Ra Radium 226.03 (2, 8, 18, 32, 18, 8, 2)	(S) 103 Lr Lawrencium (260) (2, 8, 18, 32, 32, 9, 2)	(N) 104 Unq Unnilquadium (261) (2, 8, 18, 32, 32, 10, 2)	(N) 105 Unp Unnilpentium (262) (2, 8, 18, 32, 32, ?, ?)	(N) 106 Unh Unnilhexium (263) (2, 8, 18, 32, 32, ?, ?)	(N) 107 Uns Unnilseptium (264) (2, 8, 18, 32, 32, ?, ?)	(N) 108 Uno Unniloctium (265) (2, 8, 18, 32, 32, ?, ?)	(N) 109 Une Unnilennium (266) (2, 8, 18, 32, 32, ?, ?)									

*Name not officially assigned

Lanthanide Series

57	58	59	60	61	62	63	64	65	66	67	68	69	70
(S) La Lanthanum 138.91 (2, 8, 18, 18, 9, 2)	(S) Ce Cerium 140.12 (2, 8, 18, 20, 8, 2)	(S) Pr Praseodymium 140.91 (2, 8, 18, 21, 8, 2)	(S) Nd Neodymium 144.24 (2, 8, 18, 22, 8, 2)	(N) Pm Promethium (145) (2, 8, 18, 23, 8, 2)	(S) Sm Samarium 150.4 (2, 8, 18, 24, 8, 2)	(S) Eu Europium 151.96 (2, 8, 18, 25, 8, 2)	(S) Gd Gadolinium 157.25 (2, 8, 18, 25, 9, 2)	(S) Tb Terbium 158.93 (2, 8, 18, 27, 8, 2)	(S) Dy Dysprosium 162.50 (2, 8, 18, 28, 8, 2)	(S) Ho Holmium 164.93 (2, 8, 18, 29, 8, 2)	(S) Er Erbium 167.26 (2, 8, 18, 30, 8, 2)	(S) Tm Thulium 168.93 (2, 8, 18, 31, 8, 2)	(S) Yb Ytterbium 173.04 (2, 8, 18, 32, 8, 2)

Actinide Series

89	90	91	92	93	94	95	96	97	98	99	100	101	102
(S) Ac Actinium (227) (2, 8, 18, 32, 18, 9, 2)	(S) Th Thorium 232.04 (2, 8, 18, 32, 18, 10, 2)	(S) Pa Protactinium 231.04 (2, 8, 18, 32, 20, 9, 2)	(S) U Uranium 238.03 (2, 8, 18, 32, 21, 9, 2)	(N) Np Neptunium 237.05 (2, 8, 18, 32, 22, 9, 2)	(N) Pu Plutonium (244) (2, 8, 18, 32, 24, 8, 2)	(N) Am Americium (243) (2, 8, 18, 32, 25, 8, 2)	(N) Cm Curium (247) (2, 8, 18, 32, 25, 9, 2)	(N) Bk Berkelium (247) (2, 8, 18, 32, 27, 8, 2)	(N) Cf Californium (251) (2, 8, 18, 32, 28, 8, 2)	(N) Es Einsteinium (254) (2, 8, 18, 32, 29, 8, 2)	(N) Fm Fermium (257) (2, 8, 18, 32, 30, 8, 2)	(N) Md Mendelevium (258) (2, 8, 18, 32, 31, 8, 2)	(N) No Nobelium (259) (2, 8, 18, 32, 32, 8, 2)

†This numbering system is used by the International Union of Pure and Applied Chemistry (IUPAC)